PHYSICS TO A DEGREE

PHYSICS TO A DEGREE

E G Thomas and D J Raine

Department of Physics and Astronomy
University of Leicester, UK

Gordon and Breach Science Publishers

Australia • Canada • France • Germany • India • Japan • Luxembourg
Malaysia • The Netherlands • Russia • Singapore • Switzerland

Property of Library
Cape Fear Comm College
Wilmington, N. C.

Copyright © 2000 OPA (Overseas Publishers Association) N.V. Published by license under the Gordon and Breach Science Publishers imprint.

All rights reserved.

No part of this book may be reproduced or utilized in any form or by any means, electronic or mechanical, including photocopying and recording, or by any information storage or retrieval system, without permission in writing from the publisher. Printed in Singapore.

Amsteldijk 166
1st Floor
1079 LH Amsterdam
The Netherlands

British Library Cataloguing in Publication Data

Thomas, E. G.
   Physics to a degree
   1. Physics – Problems, exercises, etc. 2. Physics –
   Examinations – Study guides
   I. Title II. Raine, D. J.
   530

ISBN: 90-5699-277-5 (softcover)

Part 1: Tutorials

Part 1: Tutorials   1

1   Energy supply and storage   1

    1.1   The Great Pyramid   1

    1.2   Energy from burning   1

    1.3   Battery power   1

    1.4   Alternative power for cars   2

    1.5   The 'Pu battery' of Cassini   2

2   Solar energy and power   2

    2.1   Energy of a burning Sun   2

    2.2   Energy of a contracting Sun   3

    2.3   Helium creation   3

    2.4   Neutrinos from the Sun   3

3   Newtonian games   4

    3.1   Athletic records   4

    3.2   Hazardous high jumps   4

    3.3   Improve your golf   4

    3.4   Snooker insights   5

4   Damage limitation   5

    4.1   Elastic and inelastic collisions   5

    4.2   Brute strength or physics?   5

5   The power of force   6

    5.1   Motion under constant power   6

    5.2   The Channel tunnel shuttle   6

    5.3   Drag racers   6

6   The centre of momentum frame   7

    6.1   A traffic accident   7

    6.2   Forbidden processes   7

    6.3   Creating the Z^0 8

7   Round and round   8

    7.1   A fairground rotor   8

    7.2   The wall of death   8

Contents

| | 7.3 | Rotating fluid | 9 |

8	About turning		9
	8.1	Walking a tightrope	9
	8.2	Tracking the angular momentum	10

9	Rocket science		11
	9.1	Steam powered space flight	12
	9.2	Single stage to orbit	12

10	Diluting gravity		13
	10.1	Rolling down inclines	13
	10.2	Galileo's legendary experiment	13
	10.3	Snooker	13

11	Gravitating bodies		14
	11.1	The great escape	14
	11.2	Solar System centre	14
	11.3	Counter-Earth	14
	11.4	Rubble piles	15
	11.5	Shuttle Dust	15

12	Oscillations		15
	12.1	Journey to the centre of the Earth	15
	12.2	Thomson's current bun	16
	12.3	Satellite attitude	16
	12.4	The expansion of solids	17

13	Resonance		18
	13.1	The opera singer and the wine glass	18
	13.2	The walls of Jericho	19
	13.3	Keep on swinging	19
	13.4	Coupled oscillators	19

14	Antigravity		21
	14.1	Cavorite	21
	14.2	Cosmological constancy	22

14.3   Bubble gravity   22

15   Orbits of the Sun   22

15.1   The wasting Sun   23

15.2   Orbital impact   23

15.3   Hale-Bopp spectacular   23

16   Accelerated frames   24

16.1   Nevertheless it moves   24

16.2   The latitude effect   24

16.3   The first battle of the Falklands   25

16.4   Accelerations of the Earth   25

17   Artificial Gravity   25

17.1   Recreational sports on Rama   26

18   Stress and strain   26

18.1   Towing ropes   27

18.2   Jumping flea   27

18.3   Into the deep   28

19   Hot and cold   28

19.1   Physics of dieting   28

19.2   Saving the *Titanic* ?   28

19.3   Winter sports   29

19.4   Warming the Earth   29

20   Cooling the Chunnel   29

21   Heat flow in ice-houses   30

22   The atmosphere   32

22.1   Columns of air   32

22.2   Climbing Everest   32

23   Atmospheric friction   33

24   Bernoulli's theorem   34

24.1   Mir   34

24.2   Aerodynamic lift   34

Contents

	24.3	The flight of a golf ball	35
25	**Spontaneous structure**		35
	25.1	Ordering life?	35
	25.2	Oil on troubled water	37
26	**Material strength**		37
	26.1	All cracked up	37
	26.2	Building in brick	38
	26.3	Up in smoke	38
	26.4	Flowing glass	38
27	**Electric shocks**		39
	27.1	Electric sight	39
	27.2	Working at high potential	40
	27.3	Electric defibrillation	40
28	**Magnetic fields**		40
	28.1	Lightning strikes	40
	28.2	Magnetic pliers	41
29	**Model circuits**		41
	29.1	The CR circuit	41
	29.2	The LCR circuit	42
30	**Dispersion on the line**		42
	30.1	Thomson's speechless cable	43
	30.2	Heaviside — speaking	43
31	**Things to do with mirrors**		44
	31.1	Solar furnaces	44
	31.2	Going to war with mirrors	45
32	**Blackbody Radiation**		45
	32.1	'Blackish' bodies	46
	32.2	Photon numbers	46
	32.3	The Sun as a blackbody	46

	32.4 The Jupiter star	47
33	Relativistic travel	47
34	Hyperbolic motion	47
	34.1 Constant acceleration in relativity	48
	34.2 Siamese rockets	48
	34.3 Rigid bodies	48
35	Mechanics near c	49
	35.1 The size of particle accelerators	49
	35.2 Relativistic snooker	49
	35.3 A relative paradox	50
36	A relativistic aberration	50
37	Lensing gravity	51
38	The Universe	53
	38.1 Olber's paradox	53
	38.2 Faster than light?	54
	38.3 Gamma-Ray bursts	55
39	The anthropic principle	55
40	Radiating gravity	56
	40.1 Dipole radiation	57
	40.2 Quadrupole radiation	58
41	Radioactive decay	58
	41.1 A natural fission reactor	58
	41.2 The Uranium clock	59
	41.3 Natural Pu	59
42	Elementary particles	59
	42.1 Catching neutrinos	59
	42.2 The long-lived proton	60
43	Quantum uncertainty	60
	43.1 An alternative Bohr	60
	43.2 Pressure Ionisation	61

	43.3	Nuclei without neutrons	61
44	**Cross sections**		61
	44.1	Supernova neutrinos	62
	44.2	Yukawa's meson	62
	44.3	Too far flung for civilisation?	63
45	**Nuclear explosions**		63
	45.1	Critical mass	64
	45.2	Nuclear fireballs	64
46	**Degenerate electron gases**		65
	46.1	A matter of compressibility	65
	46.2	Brown Dwarfs	65
	46.3	White Dwarfs	66

Part 2: Answers to Tutorials

			67
1	**Energy supply and storage**		67
	1.1	The Great Pyramid	67
	1.2	Energy from burning	67
	1.3	Battery power	68
	1.4	Alternative power for cars	68
	1.5	The Pu battery of Cassini	69
2	**Solar energy and power**		69
	2.1	Energy of a burning Sun	69
	2.2	Energy of a contracting Sun	70
	2.3	Helium creation	70
	2.4	Neutrinos from the Sun	70
3	**Newtonian games**		71
	3.1	Athletic records	71
	3.2	Hazardous high jumps	72
	3.3	Improve your golf	73
	3.4	Snooker insights	74
4	**Damage limitation**		75
	4.1	Elastic and inelastic collisions	75

	4.2	Brute strength or physics?	76
5	**The power of force**		76
	5.1	Motion under constant power	76
	5.2	The Channel tunnel shuttle	77
	5.3	Drag racers	78
6	**The centre of momentum frame**		79
	6.1	A traffic accident	79
	6.2	Forbidden processes	79
	6.3	Creating the Z^0	79
7	**Round and round**		82
	7.1	A fairground rotor	82
	7.2	The wall of death	82
	7.3	Rotating fluid	83
8	**About turning**		84
	8.1	Walking a tightrope	84
	8.2	Tracking the angular momentum	86
9	**Rocket science**		87
	9.1	Steam powered space flight	87
	9.2	Single stage to orbit	89
10	**Diluting gravity**		90
	10.1	Rolling down inclines	90
	10.2	Galileo's legendary experiment	91
	10.3	Snooker	92
11	**Gravitating bodies**		94
	11.1	The great escape	94
	11.2	Solar System centre	94
	11.3	Counter-Earth	95
	11.4	Rubble piles	97
	11.5	Shuttle dust	98
12	**Oscillations**		99
	12.1	Journey to the centre of the Earth	99

	12.2	Thomson's current bun	100
	12.3	Satellite attitude	101
	12.4	Non-harmonic forces	102
13	Resonance		105
	13.1	The opera singer and the wine glass	105
	13.2	The walls of Jericho	106
	13.3	Keep on swinging	106
	13.4	Coupled oscillators	107
14	Antigravity		109
	14.1	Cavorite	109
	14.2	Cosmological constancy	110
	14.3	Bubble gravity	110
15	Orbits of the Sun		111
	15.1	The wasting Sun	111
	15.2	Orbital impact	112
	15.3	Hale-Bopp spectacular	113
16	Accelerated frames		115
	16.1	Nevertheless it moves	115
	16.2	The latitude effect	118
	16.3	The first battle of the Falklands	119
	16.4	Accelerations of the Earth	119
17	Artificial gravity		119
18	Stress and strain		123
	18.1	Towing ropes	123
	18.2	Jumping flea	124
	18.3	Into the deep	124
19	Hot and cold		125
	19.1	Physics of dieting	125
	19.2	Saving the *Titanic?*	126
	19.3	Winter sports	126

Contents

	19.4 Warming the Earth	127
20	Cooling the chunnel	128
21	Heat flow in ice-houses	129
22	Atmospheres	131
	22.1 Columns of air	131
	22.2 Climbing Everest	132
23	Atmospheric friction	133
24	Bernoulli's theorem	136
	24.1 Mir	136
	24.2 Aerodynamic lift	137
	24.3 The flight of a golf ball	138
25	Spontaneous structure	139
	25.1 Ordering life?	139
	25.2 Oil on troubled water	140
26	Material strength	141
	26.1 All cracked up	141
	26.2 Building in brick	142
	26.3 Up in smoke	142
	26.4 Flowing glass	142
27	Electric shocks	144
	27.1 Electric sight	144
	27.2 Working at high potential	145
	27.3 Electric defibrillation	146
28	Magnetic fields	146
	28.1 Lightning strikes	146
	28.2 Magnetic pliers	147
29	Model circuits	148
	29.1 The CR circuit	148
	29.2 The LCR circuit	149
30	Dispersion on the line	151
	30.1 Thomson's speechless cable	151

Contents

	30.2 Heaviside - speaking	151
31	Things to do with mirrors	153
	31.1 Solar furnaces	153
	31.2 Going to war with mirrors	154
32	Blackbody radiation	155
	32.1 'Blackish bodies'	155
	32.2 Photon numbers	157
	32.3 The Sun as a blackbody	157
	32.4 The Jupiter star	158
33	Relativistic travel	159
34	Relativistic hyperbolic motion	160
	34.1 Constant acceleration in relativity	160
	34.2 Siamese rockets	160
	34.3 Rigid bodies	161
35	Mechanics near c	162
	35.1 The size of particle accelerators	162
	35.2 Relativistic snooker	163
	35.3 A relative paradox	164
36	A relativistic aberration	165
37	Gravitational Lensing	168
38	The Universe	170
	38.1 Olber's paradox	170
	38.2 Faster than light?	172
	38.3 Gamma-ray bursts	174
39	The anthropic principle	174
40	Radiating gravity	176
	40.1 Dipole radiation	176
	40.2 Quadrupole radiation	178
41	Radioactive decay	179
	41.1 A natural fission reactor	179
	41.2 The Uranium clock	180

41.3   Natural Pu   180

42   Elementary particles   180

42.1   Catching neutrinos   180

42.2   The long-lived proton   181

43   Quantum uncertainty   182

43.1   An alternative Bohr   182

43.2   Pressure ionisation   182

43.3   Nuclei without neutrons   184

44   Cross sections   185

44.1   Supernova neutrinos   185

44.2   Yukawa's meson   186

44.3   Too far flung for civilisation   187

45   Nuclear explosions   188

45.1   Critical mass   188

45.2   Nuclear fireballs   189

46   Degenerate electron gases   190

46.1   A matter of compressibility   190

46.2   Brown dwarfs   191

46.3   White dwarfs   192

Part 3: General Physics Problems   193

1   Energetics   193

2   Mechanics in action   193

3   Power in space   194

4   Tethered satellites   195

5   The fate of the Earth   196

6   Perfect rockets   196

7   The speed limit for wheel-driven vehicles   197

8   The universality of free-fall   197

9   Bubbles   198

10   The deep blue sea   198

11   All hail   198

12   Escape into space   198

Contents

13	In orbit	199
14	Journey to Mercury	199
15	Not so square	200
16	The tractor beam of the *Enterprise*	200
17	Orbits inside matter	201
18	The Indian rope trick	202
19	Time and tides	203
20	The day the Earth caught fire	204
21	Space debris	205
22	Disposing of space junk	205
23	Suspension bridges	206
24	Tennis then and now	207
25	Energy in life	207
26	Hot potatoes	208
27	Thermal balance in mammals	209
28	Over(h)eating	209
29	Dinosaur metabolism	210
30	Dropping like lead	210
31	Journey to the centre of the Earth	211
32	Buoyancy	212
33	Survival on Venus	212
34	The Gulf Stream	213
35	The Antarctic ice cap	213
36	C_d's	213
37	Fluid flows	214
38	Land yachts	215
39	Bouncing bombs	215
40	Mirrors without the grind	216
41	Sand and water	217
42	Thunder and lightning	217
43	Diverging beams	218
44	A poor man's black hole	218
45	The Ur-Star	219

46	Hearing things	219
47	Intensity of sound	220
48	Faster than sound	220
49	Viewing the Sun	220
50	Laser links	221
51	The laser pen is mightier	221
52	The Green Flash	222
53	Dispersion	222
54	Focussing neutrons	223
55	Resolution	223
56	The Earthlit Moon	223
57	A natural hazard	224
58	Fusion power	225
59	Cold fusion	225
60	Boson stars	225
61	The Landau atom	226
62	Colourful solids	227

Physical Constants

elementary charge	e	1.602×10^{-19} C
electron rest mass	m_e	9.109×10^{-31} kg
proton rest mass	m_p	1.673×10^{-27} kg
neutron rest mass	m_n	1.675×10^{-27} kg
Planck constant	$h = 2\pi\hbar$	6.626×10^{-34} Js
Gravitation constant	G	6.673×10^{-11} N m^{2}kg^{-2}
Speed of light in vacuum	c	2.998×10^8 m s^{-1}
Boltzmann constant	k	1.381×10^{-23} J K^{-1}
Stefan constant	$\sigma = \frac{ac}{4}$	5.671×10^{-8} W m^{-2}K^{-4}
Radiation constant	$a = \frac{4\sigma}{c}$	7.564×10^{-16} J m^{-3}K^{-4}
Avogadro number	N_A	6.022×10^{23} (gm mole)$^{-1}$
		6.022×10^{26} (kg mole)$^{-1}$
Permittivity of vacuum	ε_0	8.854×10^{-12} F m^{-1}
Permeability of vacuum	μ_0	$4\pi \times 10^{-7}$ H m^{-1}
atomic mass unit	u	1.661×10^{-27} kg
		931.494 MeV
Bohr magneton	μ_B	
1eV		1.602×10^{-19} J
Classical electron radius	$r_0 = \frac{e^2}{4\pi\varepsilon_0 mc^2}$	2.818×10^{-15} m
Acceleration due to gravity	g	9.81 m s^{-1}
Atmospheric pressure at sea level		1.01×10^5 N m^{-2}
Specific heat of water		4190 J kg^{-1} K^{-1}

Astronomical Constants

Solar Mass	$M_\odot$	1.989×10^{30} kg
Solar Radius	$R_\odot$	6.960×10^5 km
Solar Luminosity	$L_\odot$	3.90×10^{26} J s^{-1}
Solar constant		1360 W m^{-2}
Age of the Sun		4.5×10^9 years
parsec	pc	3.086×10^{16} m
Mean Earth-Sun distance	1AU	1.496×10^{11} m
Mean Earth Moon distance		3.84×10^5 km
Earth Mass		5.97×10^{24} kg
Moon Mass		7.35×10^{22} kg
Mean radius of Earth		6370 km
Seconds in a year		3.156×10^7

Preface

First perhaps we should explain the title. The book is not, as one reviewer thought, about physics at the 'A-degree' level (which does not exist). The book is about certain aspects of undergraduate degree level physics. It concerns problems that can be solved using conceptual insight, a knowledge of basic theory, and mathematical techniques no harder than the basic calculus of several variables. This is precisely the sort of problem-solving that is begun as part of an undergraduate physics degree programme, but which is often not developed to the point where it gets interesting. Our book is designed to fill this gap.

To a degree then, this book is about making the transition from knowing physics to doing physics. In UK universities it is this that is tested by 'general physics' examinations, but, until recently, was perhaps less widely taught there. It was assumed that students would pick up these skills without explicit instruction. This approach has been criticised by employers who complain that physics graduates would often be able to do a variety of impressive things, such as writing out from memory the solution of Schrödinger's equation for the hydrogen atom, yet would be unable to use a little physical insight and some elementary grade physics to solve a real-life problem. Our aim, therefore, is to provide material from which students can learn the skills which physicists need, but which are difficult to inculcate in the context of topics-based course work. We are not attempting to teach core material through examples; in this we differ from apparently similar books of problems. In particular we have not attempted to balance questions between areas of core physics. We also differ from others in the range of difficulty and level of the problems which span a UK undergraduate degree programme.

It is not obvious what makes a good general physics question. It is certainly something more than the attempt to achieve relevance by having a car driven by Harry crash into one driven by Sally. In fact it probably has little to do with 'relevance'. Critics who cannot see (to take an headline example in the British press from a few years ago) that to construct the field of view of a goldfish is a good pedagogical undergraduate project in physics, misunderstand physics, relevance, pedagogy and probably also goldfish. Problems should at least be sufficiently intriguing to make the student want to work them out. Thus, the student should feel it spoils the fun to turn to the back of the book to look up the answer. That having been said, we hope that those who do just that will find it an entertaining way to read the book.

In Part 1 of the book the problems are arranged into short groups that we have called 'tutorials' (for want of a better word). We believe that many of these can be used as the basis of a tutorial (and we have done so with some, but not all).

But that is not their sole function. In fact, most of the book could be worked through by a student on their own or used for group problem sessions. We have given detailed answers to these tutorials in Part 2. It is important to remember though that throughout the book we assume the reader has an initial familiarity with the relevant material. One test is that it should not be necessary to look up the meaning of the symbols for standard physical quantities in a problem. Within this context we have tried to make the answers as complete as we reasonably can. In particular, we have given the substitution of numerical values into equations explicitly in 'engineering format' (numbers and units) to make it easy to follow the calculation of a numerical answer.

In Part 3 we present some problems, without answers, of the sort that we use at Leicester in group work for third year undergraduates. For the benefit of our staff supervisors we do have answers, and these are available, from the authors (not the publisher) to Physics Departments (not individuals) who adopt the book for undergraduate use.

Not all of our problems are original, probably not even many of those that we believe to be so. We have acknowledged borrowings where we can recall them. In some cases, we have simply taken well known problems and put a little 'spin' on them. We are grateful to our colleagues Dr Richard Jameson and Dr Tony Janes, who volunteered some ideas, to Drs Steve Gurman, Gordon Stewart, Richard Willingale and Professor Ted Davis for various suggestions, and for specific help on Barnes Wallis and bouncing bombs from The Brooklands Museum and Mr N W Boorer, on cold fusion from Dr George Fraser, and on the Chicxulub meteorite impact from Dr Peter Maguire. Tom Preston and the late Professor Aisa Blakely supplied data. We hope our acknowledgements are complete.

Part 1: Tutorials

Tutorial 1   Energy supply and storage

1.1   The Great Pyramid

The Greek historian Herodotus reported that it took the labour of 100,000 men to build the Pyramid of Khufu at Giza. Since he was writing 2000 years after the event we cannot place much reliance on this figure. However, we can use physics to assess the accuracy of the account. As the pyramid was presumably built using human muscle power alone, it is possible to estimate the labour that went into its construction from the amount of potential energy that is stored in the pyramid (*Nature*, **383**, 218, 1996) and the fact that a man can perform about 2.5×10^5 J of useful work in a day.

(a) Estimate the minimum number of man days needed to lift the stones into place given that the pyramid had a height at that time of 146.7 m and that its base is a square of side 230.4 m. Take the density of stone to be 2700 kg m^{-3}.

(b) Given that an upper limit to the time available for the construction is the length of Khufu's reign, which was 23 years, estimate the size of workforce needed for the task.

1.2   Energy from burning

(a) What is the specific energy storage capacity of coal in J kg^{-1} (i) excluding the mass of oxygen needed to burn it and (ii) including the oxygen? In the combustion of coal each atom of carbon combines with one molecule of oxygen yielding about 4 eV.

(b) Use your result to estimate the energy yield per unit weight of a high explosive.

1.3   Battery power

(a) How much energy is stored in a 1.5V long life alkaline battery, given that two such batteries in series will power a 2W torch bulb, under favourable conditions, for 27 hours?

(b) It is interesting to work out the cost per kWhr of electricity bought in this form. At the time of writing such a battery retails in the U.K. for about £1.50. Do the calculation and compare the result with the cost per kWhr (£0.08 currently) charged by electricity companies for mains electricity.

1.4 Alternative power for cars

(a) What is the specific energy storage capacity in J kg^{-1} of a 12 V car battery having a capacity of 60 ampere hours and a weight of 12 kg?

(b) What weight of batteries would be needed to propel a car the same distance as a tank (40 litres) of petrol? The energy yield from burning hydrocarbons is about 50% greater than that from coal.

(c) How much power could solar energy supply to a car under favourable circumstances?

1.5 The 'Pu battery' of Cassini

(a) The Cassini spacecraft launched in October 1997 on a voyage to Saturn is powered by 33 kg of ^{238}Pu. Estimate the energy released by the decay of 1 kg of ^{238}Pu. (^{238}Pu emits an alpha particle of energy 5.5 MeV.)

(b) Although the energy storage capacity of ^{238}Pu is enormous, as the calculation shows, the rate at which it can be extracted is fixed by the radioactive decay law. How much power will be available to the spacecraft on its arrival at Saturn in 2004?

The half life of ^{238}Pu is 88 years.

Tutorial 2   Solar energy and power

We investigate the source of the Sun's energy, a problem that exercised nineteenth century physics (once the conservation of energy was established) but was not resolved until the 1930s, when nuclear reactions were first understood.

2.1 Energy of a burning Sun

(a) Estimate the energy output of the Sun over its lifetime of 4.5×10^9 years assuming that its luminosity has not changed over time. The solar constant is 1360 W m^{-2} and the distance of the Earth from the Sun is 1.5×10^8 km.

(b) Show that we cannot account for this energy output by making the assumption that the Sun is composed of best quality coal and a supply of oxygen (see question 1.2). Why does this calculation also rule out other chemical processes as the source of the Sun's energy ?

2.2 Energy of a contracting Sun

Heinrich Helmholtz (Popular Scientific Lectures, 1908, **2**, 312-15), calculated the lifetime of the Sun assuming the source of the Sun's luminosity is gravitational energy released in the course of the Sun's contraction from a cloud of gas to its present size, an idea independently suggested by Lord Kelvin.

(a) Estimate the energy released by the contraction of the Sun from a dispersed state to its present diameter assuming that the density remains uniform during contraction.

(b) What would be the lifetime of the Sun radiating at its present rate according to this hypothesis?

We can in fact increase the energy yield from this process to the value obtained in part (a) but only by concentrating the Sun's mass towards its centre to a physically unrealistic extent. The short time scale was in conflict with the beliefs of geologists and evolutionists at the time. Note that the source of the Sun's energy supply became a problem only in the nineteenth century with the discovery of the conservation of energy.

2.3 Helium creation

The actual mechanism by which energy is generated in the Sun is nuclear fusion of protons into helium 4, a process which releases 26.7 MeV of energy per helium nucleus formed. Estimate the amount of helium which has been created by this process over the Sun's lifetime.

Observation tells us that the Sun and other stars contain about 25% by mass of helium 4. So we must conclude that most of the helium present in the Sun and stars was present in the material from which they formed. The universal presence of this helium of non-stellar origin is strong evidence in support of the hot big-bang cosmological model because this model predicts the production of helium in the early universe.

2.4 Neutrinos from the Sun

In addition to the electromagnetic radiation coming from its surface the Sun emits neutrinos from its centre. For each helium atom formed, two neutrinos are emitted with an average energy of about 0.3 MeV.

(a) Estimate (i) the energy flux of solar neutrinos at the Earth and (ii) the energy per second passing through your body.

(b) How much difference does the inclusion of the energy flux in neutrinos make to the estimate of helium produced by the Sun? (Question 2.3.)

Tutorial 3   Newtonian games

The following examples illustrate the use of conservation laws in dynamical problems in which there are no dissipative forces. In this tutorial the relevant laws are conservation of mechanical energy and conservation of linear momentum. Conservation laws apply to isolated systems and are useful when motion is transferred from one part of the system to another. In question 3.3 the system is not strictly isolated but we can treat it as such to a sufficient approximation. The moral is simple: dynamics problems are difficult; always begin by asking if anything is (or can be considered to be) conserved.

3.1   Athletic records

The year is 2100 and the pole vault record stands at 7.5 metres. Estimate the world record for the 100 metres sprint at this time.

This question is one of our favourites since it shows the power of physics to make the apparently most unlikely connections. Incidentally, both records would be held by women by this date – see Nature, Jan 2 1992.

3.2   Hazardous high jumps

What is the minimum radius of a spherical asteroid on which it would be safe to hold the inter-planetary Olympic high jump (or long jump i.e. from which an earthling cannot jump to infinity)? You may assume that the density of an asteroid is 2×10^3 kg m^{-3}. The world high jump record is currently 2.4 m.

3.3   Improve your golf

(a) When a golfer strikes a golf ball the ball and clubhead are in contact for a time of about 0.5 ms. The shortness of this time of contact means that linear momentum of the ball and clubhead is conserved during collision despite the fact that the clubhead is attached via the clubshaft to the golfer. Explain this.

(b) Why is the contact time between the ball and club largely independent of the strength of the blow?

(c) A golfer wants to increase the length of his drive. Should he (i) endeavour to swing the club say 10% faster or (ii) choose a club with 10% more mass in its head and try to maintain his usual clubhead speed?

(d) Would using a club that is 10% longer in the shaft be a better solution than either of the above strategems?

The mass of a golf ball is 0.046 kg and the typical mass of a golf club head is 0.2 kg.

3.4   Snooker insights

A useful rule of thumb for snooker players is that after a moving ball has made a glancing collision with a stationary ball the angle between their diverging paths is 90^0.

(a) Show that this rule is a consequence of the conservation of linear momentum and conservation of energy.

(b) Why is the rule likely to be violated to some extent in practice?

Tutorial 4   Damage limitation

In the ideal case of a perfectly elastic collision between two bodies no damage is done to the bodies since they do not absorb any energy, except temporarily and reversibly. Damage to a body requires irreversible expenditure of energy, so the collisions which cause the damage must be inelastic. The extent to which a collision is inelastic depends on the materials and forces involved. Exceeding the elastic limit of the material produces permanent deformation and exceeding the breaking strain results in rupture.

4.1   Elastic and inelastic collisions

(a) When a golfer strikes a golf ball no damage is done to the ball or the club, yet the collision is not perfectly elastic. Explain where the missing energy has gone.

(b) Rubber and aluminium bullets of the same size and mass are fired at a wooden block as target. Which is more likely to i) knock the target over; ii) damage the target?

(c) Which rugby tackle is more effective: one by a scrum-half or one by a prop-forward if both have the same momentum? Which is likely to be more painful for the player tackled? [For non-followers of the game, the prop-forward is the more massive.]

4.2   Brute strength or physics?

In days gone by the strongman at the circus would perform the following feat. He would lie on the ground and allow a heavy slab to be placed on his chest. A member of the audience was invited to strike the slab as hard as he could with a sledge hammer. The strongman would emerge from this ordeal with all his ribs intact. Explain how this is possible.

Tutorial 5   The power of force

A body subject to a constant force undergoes constant acceleration. However, in practice, it is often not the force that is constant but the power being supplied (i.e. the rate of doing work is constant). A constant supply of power to a moving body does not generally give it a constant acceleration.

5.1   Motion under constant power

(a) If power P is supplied to a body of constant mass moving at speed v, what is the force acting on the body?

(b) Is the rate of change of momentum constant?

(c) Is the rate of change of energy constant?

(d) At constant acceleration v increases linearly with time. Show that at constant power v^2 is linear in time.

5.2   The Channel tunnel shuttle

The shuttle trains that run from Dover to Calais through the Channel tunnel have at total weight of 2500 tonnes and can draw 11 MW of electrical power. Assume that the power drawn is a constant 11 MW whilst the train is accelerating from rest up to its running speed of 140 kph and ignore frictional loses. (In fact, accelerating from rest at constant power is an unrealistic assumption. It would require an infinite initial acceleration hence an infinite initial force. The initial force is limited in practice by wheel spin.)

(a) Determine how the acceleration of the train depends on its velocity.

(b) Determine how the velocity depends on time and hence the time taken to reach the running speed.

(c) Determine how the velocity depends on distance. Hence what is the distance needed to reach the running speed?

(d) Compare the above quantities for a train which accelerates under a constant force that is equal to the time-average of the real force.

5.3   Drag racers

In a drag race the competing cars accelerate from rest over a straight course a quarter of a mile long and the one with the highest terminal speed wins. R. Huntingdon (*Am. J. Phys.*, 1973, **41**, 311) discovered the empirical law that

terminal velocity $\propto$ (maximum power output of the car)$^{1/3}$.

Show that this rule is consistent with acceleration under constant power.

Tutorial 6   The centre of momentum frame

The message of the principle of relativity, whether it is Newtonian or Einsteinian relativity, is that all inertial frames are equivalent as far as the laws of mechanics are concerned. However, in practice, many problems can be solved more readily by a judicious choice of frame. In collision problems the best choice is usually the centre of momentum frame. This is the frame in which the total momentum of the system is zero. In this tutorial we illustrate the use of the centre of momentum frame. Some useful background is given in tutorial 4.

6.1   A traffic accident

Consider the following two collisions between a pair of cars.

Collision (i): A car travelling at 60 mph (96 km h^{-1}) collides with a stationary car.

Collision (ii): Two cars each travelling at 30 mph (48 km h^{-1}) collide head on.

It is clear that the single moving car in (i) has more kinetic energy than the two moving cars in (ii). But the need to conserve momentum in collision (i) means that not all the kinetic energy is available to cause damage, whereas in collision (ii) all the kinetic energy can go into causing damage and momentum will still be conserved. So the message is clear: to determine the energy that can be made available we have to transfer to the centre of momentum frame.

Bearing this in mind which of the above collisions, assuming they are both inelastic, causes the most damage?

6.2   Forbidden processes

(a) The photoelectric effect is the absorption of a photon by an electron: it is responsible for ejecting electrons from an illuminated metal surface and for the ionisation of atoms. However photoelectric absorption by isolated electrons is never observed. Why not?

(b) How are energy and momentum conserved in the ionisation of an atom, which can take place in isolation?

(c) The production of an electron-positron pair from a photon can occur when the photon energy exceeds the combined rest energy of the pair. The process is observed when gamma rays interact with matter. Pair production by isolated gamma rays is never observed. Explain why this is.

6.3   Creating the Z^0

(a) The Super Proton Synchrotron at CERN was originally conceived to accelerate protons to an energy of 500 GeV and then to collide them with protons in a stationary target. What is the maximum energy available for creating new particles in the collision of a 500 GeV proton with a stationary proton ? Hint: Use the energy-momentum transformation of special relativity.

(b) This energy falls well short of the rest energy of the Z^0 boson which is 91 GeV. What beam energy would be required to produce Z^0 particles?

The Z^0 particle is a prediction of the theory that unites electromagnetic and weak interactions. In 1983 an ingenious modification of the SPS was carried out which enabled Z^0s to be produced without any increase in the beam energy. The SPS was converted into a collider by injecting bunches of anti-protons into the same ring as the protons. With such an arrangement the anti-protons are accelerated to the same peak energy as the protons but move in the opposite sense around the ring. Colliding the protons with the antiprotons gives a centre of mass energy which is twice the beam energy rather than the fraction of it that the old mode of operation of the SPS gave.

Tutorial 7   Round and round

7.1   A fairground rotor

A fairground rotor consists of a cylindrical room which can be spun about its axis of symmetry. Intrepid members of the public are invited to stand inside the room with their backs touching the wall and their feet on the floor whilst the room is spun up. When the room has reached a certain angular velocity the floor is lowered and the people inside find themselves stuck to the wall.

(a) How fast does a 4 m diameter rotor have to be spinning before the floor can be lowered?

(b) The minimum angular velocity at which it is safe to lower the floor depends somewhat on the size of the individual. Explain why this is so. (If you do not see this follows from the first part of the question, examine your assumptions!)

7.2   The wall of death

The diagram (figure 1) shows a motorcyclist riding on the 'wall of death' (the inside surface of a cylinder of radius 5 m).

(a) What is the minimum speed that the motorcyclist can have on the wall?

(b) How does the angle θ of the bike depend on its speed?

(c) What is the maximum angle θ that the bike can have?

(d) If the motorcyclist takes a pillion passenger how does this affect the answers to the above questions?

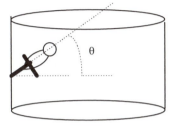

Figure 1. A rider on the wall of death

7.3   Rotating fluid

Exactly the same ideas can be used to determine the shape of the surface of a rotating fluid. Show that the surface of a cup of tea that has been stirred is a paraboloid.

This looks like a problem in fluid mechanics (and can indeed be solved as a problem in hydrostatics). But since the fluid appears at rest in a rotating frame, the fluid properties cannot matter. It is therefore possible to treat each particle of the fluid separately, and hence to solve the problem using particle mechanics.

Tutorial 8   About turning

8.1   Walking a tightrope

Tightrope walkers usually perform their feats with the aid of a long balancing pole. The idea is that any imbalance can be corrected by moving the pole a small horizontal distance to create a counter turning moment. It also increases the walker's moment of inertia which reduces the acceleration under a gravitational torque and provides more time for corrective action. If the pole is flexible an additional benefit is that it lowers the centre of gravity of the walker.

(a) Let the walker have mass M and height l and let the pole have mass m and length L,and assume that the mass per unit length of the pole and the walker is uniform. Let G be the torque on the system due to gravity when the axis of the system is displaced through an angle θ from the vertical and let fl be the height of the centre of gravity of the pole above the wire (see figure 2). For a rigid pole $f \approx 1/2$ and for one that droops at the ends $f < 1/2$. Show that the equation of motion of the system (walker and rigidly held pole) is

$$I\ddot{\theta} = G$$

where, the moment of inertia I is given by $I = \frac{1}{3}Ml^2 + \frac{1}{12}mL^2 + m(fl)^2$

(b) Suppose that the walker starts to tilt away from the vertical. Show that after a small time t the angle of tilt is given by

$$\theta \simeq \Lambda t$$

where Λ^{-1} is the time constant of the system.

(c) To determine how effective the pole is in slowing up the rate of toppling evaluate the ratio of the time constant for the walker with the pole to the time constant without the pole. Take $L = 12$ m and $m = 14$ kg which are the upper limits to the dimensions of balancing poles in use today.

(d) The time constant can be further increased by carrying a non-uniform pole. Evaluate the ratio of time constants in (c) for a pole which has its weight concentrated at its ends.

(e) What is the ideal physique for a tightrope walker?

You can obtain more information on the internet by searching under funambulism.

8.2   Tracking the angular momentum

(a) The Sun has over 99% of the mass of the Solar System but less than 1% of the angular momentum. Very young stars are observed to be rotating rapidly with

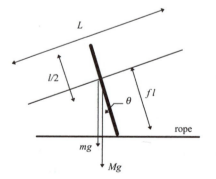

Figure 2. Mass model for the tightrope walker and pole

rotational velocities of about 200 km s^{-1}. Estimate roughly the angular momentum of the Young Sun and compare it with the present angular momentum. (The present Sun rotates once every 27 days.)

(b) It is believed that a stellar wind (outflow of matter from the surface to infinity) is responsible for taking away the angular momentum of a star in its early life. The wind from a young star has approximately 1000 times the mass flux of the wind from the present day Sun. The wind particles follow radial magnetic field lines in the co-rotating frame out to a distance of about 1 a.u. where they decouple from the star. If this young phase lasts 10^8 years, show that the solar wind can carry away most of the initial angular momentum without appreciably affecting the mass.

The present mass flux in the Solar wind is about $2 \times 10^{-14} M_\odot$ yr^{-1}.

Tutorial 9   Rocket science

In many science fiction stories written in the 1950s, the Dan Dare adventures for example, space travel is accomplished in single-stage, rocket-propelled spacecraft which can take off from the ground and travel to some other planetary destination in the solar system without refuelling en route. The object of this tutorial is to understand why we still have not got Dan Dare's technology and why in all probability we never will have.

Chemically powered rockets are the only known means of getting from the Earth's surface into space. Their performance is governed by the rocket equation which for vertical upwards motion in a gravitational field is

$$m\frac{du}{dt} = -v_0 \frac{dm}{dt} - mg.$$

Integrating with respect to time t, assuming $dm/dt = -\dot{m} =$ constant, gives

$$u = v_0 \log_e(m_0/m) - g(m_0 - m)/\dot{m},$$

where m_0 is the take off mass, m the mass when the fuel has been burnt, v_0 the exhaust velocity, i.e. the velocity at which the exhaust gases leave the rocket, and $\dot{m}$ the rate at which fuel is expelled from the back of the rocket. Air resistance in the lower atmosphere has been ignored as we are interested here in finding an upper limit to rocket performance. The rocket equation tells us that to achieve this maximum performance we need to (i) choose a fuel which gives the highest possible exhaust velocity, (ii) maximise the mass ratio m_0/m by constructing the rocket out of the lightest materials consistent with the requirements of structural strength and (iii) develop highly efficient motors to maximise the quantity $\dot{m}$.

9.1   Steam powered space flight

Rockets powered by liquid hydrogen and liquid oxygen get their energy from the reaction

$$2H_2 + O_2 = 2H_2O,$$

which gives 13.3MJ kg^{-1} of fuel combusted - the highest value from any rocket fuel in current use. The energy appears as thermal energy of the combustion product (steam) but the efficiency of conversion to directed kinetic energy, de-fined as the ratio of kinetic energy of the exhaust gases to the chemical energy released by the fuel, cannot be 100%.

(a) Explain why this is and decide where the waste heat goes.

(b) Assume that the hot gas, which can be taken to be ideal, starts from rest in the combustion chamber and expands adiabatically to reach atmospheric pressure at the exit of the rocket nozzle. Show that the efficiency of an ideal rocket is the same as that of a Carnot engine.

(c) Estimate the efficiency of an ideal rocket motor in which the pressure in the combustion chamber is 7.0 MPa.

(d) Obtain an estimate of the exhaust velocity of this fuel.

In practice, for the purpose of deriving the correct thrust, an 'equivalent exhaust velocity' is often quoted in data tables which is slightly larger than the actual exhaust velocity. The equivalent exhaust velocity of the Space Shuttle's main engine is about 4500 m s^{-1}. In employing the rocket equation the equivalent exhaust velocity should be used for v_0 because this includes the pressure thrust from the ejected gas in addition to its momentum thrust.

9.2   Single stage to orbit

Design studies for a single stage reusable rocket powered vehicle (the Delta Clip-per) have concluded that it is possible to achieve a mass ratio m_0/m of about 10 by using modern light weight materials. Rocket motors can give a thrust to launch weight ratio between 1.5 and 2. Show that these parameters would al-low a single stage rocket to reach low Earth orbit but not enable it to achieve escape velocity from the Earth. This shows that Dan Dare's exploits remain in the realms of science fiction.

Once in space it is possible to use low thrust ion engines. These engines de-velop low power but can maintain this for long periods of time, the converse of chemical engines which generate high power for short periods.

Tutorial 10   Diluting gravity

This tutorial concerns Galileo's test of the independence of the gravitational acceleration of a body on its composition. To appreciate it requires some familiarity with objects rolled down inclined planes which we address first.

10.1   Rolling down inclines

(a) A circular disc of mass M and radius a rolls without slipping with velocity v. What is the linear momentum of the disc.

(b) What is the angular momentum of an annulus of the disc between radii r and $r+dr$. Hence, what is the total angular momentum of the disc?

(c) What is the total energy of the rolling disc?

(d) What are the answers to the corresponding questions if the disc is replaced by a sphere?

(e) How, qualitatively, are the above results changed if the disc is slipping?

10.2   Galileo's legendary experiment

Galileo is credited with establishing that all bodies fall with the same acceleration under gravity. He did this not by dropping bodies from the leaning tower of Pisa, as legend has it, but by rolling balls down an inclined plane. This has the advantage of diluting gravity which makes it easier to measure the time of fall. However, Galileo was fortunate in the shapes of bodies he chose to compare.

(a) Show that a sphere of lead and a sphere of wood rolling down an inclined plane cover the same distance in the same time. [Hint: Write down the conservation of energy for a rolling ball and differentiate it with respect to time to find the acceleration.]

(b) Does the size of the balls make any difference?

(c) Does the shape of the rolling object make any difference?

(d) Would Galileo have made his discovery if he had compared the rolling motion of spheres and cylinders?

10.3   Snooker

We can apply these ideas to the game of snooker. When a snooker ball is struck with a cue at its mid-point it slides before rolling. To eliminate the initial sliding phase the ball must be struck at a point above its centre.

(a) Show that this point is $7/10^{ths}$ the height of the ball above the table and does not depend on the strength of the blow. [Hint: an off-centre blow imparts a torque as well as a translational force; impulse = change in linear momentum; impulsive torque = change in angular momentum.]

(b) At what height should the side cushions be above the table top?

(c) Would the 7/10^{ths} rule apply if the game were to be played on the Moon (i.e. if we were to dilute gravity)?

(d) Nevertheless the game of snooker would be slightly different on the Moon. Explain why.

(e) Prior to a match between two good snooker players a practical joker replaces the normal snooker balls by balls having the correct size and weight but with a density that increases radially towards the centre. Explain why the players would experience a loss of their usual form.

Tutorial 11   Gravitating bodies

11.1   The great escape

(a) Show that the gravitational force exerted by the Sun on the Moon is about twice as great as the gravitational force exerted by the Earth on the Moon.

(b) Why does the Moon not escape from the Earth?

11.2   Solar System centre

Does the centre of mass of the Solar System ever lie outside the Sun? The radius of the Sun is 696×10^3 km. The masses of Jupiter, Saturn and Neptune, the three largest planets, are respectively 19.0×10^{26} kg, 5.7×10^{26} kg and 1.0×10^{26} kg and their distances from the Sun are respectively 7.8×10^8 km, 14.3×10^8 km and 45.0×10^8 km.

11.3   Counter-Earth

According to a conjecture of the Pythagorians the Solar system contains not only the Earth but also a counter-Earth moving in the same orbit but always hidden behind the Sun.

(a) Supposing that the counter-Earth has the same mass as the Earth, by how much would its presence alter the length of the Earth year?

(b) If there were a counter-Earth would it always be hidden behind the Sun?

Note that although this is a three body problem it is an example that can be solved analytically.

11.4 Rubble piles

It has been suggested (*Science*, 1996, **272**) that some asteroids might be flying rubble piles, that is to say they consist of a loose agglomeration of pieces of rock held together only by gravity. The asteroid Matilda is believed to be an example. Show that if this is true the minimum rotation period that such an asteroid could have is about 2.3 hours irrespective of its size. Take the density of asteroid rock to be 2000 kg m^{-3}.

11.5 Shuttle Dust

In Earth orbit the Space Shuttle is not a perfect inertial frame and observations carried out over a sufficient length of time will reveal this. Estimate this length of time by considering a particle of dust which is deposited 1 m further from the Earth than the centre of mass of the Shuttle. How long would the particle take to drift to the shuttle wall a distance of, say, 1 m?

Tutorial 12   Oscillations

Simple harmonic motion occurs whenever the displacement of a body from a position of equilibrium gives rise to a harmonic restoring force, i.e. a restoring force which is proportional to the displacement. In the absence of dissipative forces, the oscillations which occur are sinusoidal with a period which is independent of the amplitude of the motion.

12.1   Journey to the centre of the Earth

(a) Suppose, like Jules Verne's heroes, you descend down the vent of an extinct volcano and make your way to the centre of the Earth. Assuming that the Earth has uniform density, express your weight in terms of your radial distance from the Earth's centre. If you were to fall freely in the vent show that your motion would be simple harmonic.

(b) Suppose now, in order to cut journey times to and from the antipodes, a hole is bored through the centre of the Earth and kept under vacuum. Travellers would board a spaceship-like vehicle secured by docking magnets at one end of the tunnel. The vehicle would then be released and would fall towards the other end of the tunnel where magnets would secure it for the passengers to disembark. How long would the journey take?

(c) How would the weight of the passengers vary over the journey?

(d) What is the maximum speed that the traveller would reach?

(e) In the absence of friction, what is the journey time to fall through a similar tunnel connecting London to Peking (which are not at the ends of a diameter)? Assume the Earth to have uniform density.

12.2   Thomson's current bun

(a) At the beginning of the twentieth century J.J.Thomson proposed a current bun model of the atom. According to this model, a hydrogen atom in its ground state consists of an electron at rest at the centre of a uniform sphere of positive charge of diameter 10^{-10}m. Find the frequency of oscillation of the electron when it is displaced from its equilibrium position.

(b) Compare the spectrum of Thomson's hydrogen atom with the actual hydrogen spectrum.

12.3   Satellite attitude

Simple harmonic motion is important in physics because, to a good approximation, it describes the motion of the multitude of systems in which the restoring force is not harmonic, provided that the oscillations have a small amplitude. To see this consider the Taylor expansion, about the system's equilibrium position, of the restoring force $F(x)$, where x is the displacement. The expansion is

$$F(x) = F(0) + F'(0)x + \frac{1}{2}F''(0)x^2 + \ldots.$$

The first term is zero, because the force vanishes at the equilibrium position, and, for small x, the second and higher order terms are small compared with the linear term.

In the following example (*Am J Phys.*, 1985, **53**, 1002) the system executes simple harmonic motion for a small displacement from equilibrium - this shows that the system is stable to small perturbations.

(a) A non-spherical satellite is in its equilibrium attitude if its 'long' axis points towards the centre of the Earth. Show this, for the case of a dumbbell satellite consisting of two equal masses separated by a light rod in Earth orbit, by demonstrating that, if the satellite is given an angular displacement from its equilibrium attitude, it will experience a restoring force, and for a small displacement will execute simple harmonic motion.

(b) Show that the period of the simple harmonic motion is independent of the mass and length of the dumbbell. How does it depend on distance from the Earth?

(c) Is the satellite stable against displacement out of the plane of the orbit?

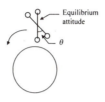

Figure 3. An oscillating dumbbell satellite (not to scale!)

12.4   The expansion of solids

The atoms in a solid are subject to a restoring force when displaced from their equilibrium position. The harmonic approximation to this restoring force, which can be obtained from the pair potential, is adequate for the treatment of bulk properties, for example the lattice specific heat, but would predict that solids do not undergo thermal expansion, which is clearly incorrect. To account for thermal expansion we need the next order anharmonic term in the expansion of the pair potential. In principle, the magnitude of the anharmonic term is a free parameter of the interatomic force. On the other hand the pair potential is completely determined by the bulk properties. The question therefore arises as to whether the pair potential gives an adequate account of thermal expansion.

(a) The potential energy of a pair of atoms in an ionic solid is given by

$$U(r - r_0) = -\frac{a}{r} + \frac{\rho_0 a}{r_0^2} e^{-(r-r_0)/\rho_0}$$

where r_0 is the equilibrium atomic separation, $\rho_0 = 0.1 r_0$ and a is a constant. Sketch the potential.

(b) Explain why this form of potential would be expected to give rise to thermal expansion.

(c) Find the Taylor expansion of $U(r - r_0)$ about $r = r_0$ to third order in $(r - r_0)$ (with coefficients in terms of the atomic mass $m = \mu m_p$, the oscillation frequency w and the separation r_0.) It is sufficient to set numerical coefficients of order unity to 1.

(d) Hence find the equation of motion to order $(r - r_0)^2$ of an atom displaced in this potential.

(e) Find the angular frequency of small oscillations w about the equilibrium in terms of the mass μm_p of the atoms of the crystal if $U_0 \ (\equiv U(r_0)) = 6$ eV and $r_0 = 0.2$ nm. What is the significance of this frequency?

(f) Show that a zero order solution of the equation of motion (neglecting the quadratic term) is $x = r - r_0 = A\sin\omega t$ and use this to derive the next order of approximation to $x(t)$.

(g) By noting that at temperature T the zero order oscillator must have energy kT deduce the coefficient of expansion of the crystal.

(h) Experimental coefficients of expansion are of order 10^{-6} K^{-1}. What do you deduce about the adequacy of the pair potential to describe anharmonic effects?

Tutorial 13   Resonance

Resonance occurs when an oscillating system is driven at its natural frequency by an applied force. The theory of the damped harmonic oscillator tells us that at resonance in steady state the energy stored in the oscillator is given by

$$E = \tau P,$$

where P is the power supplied by the driving force and $\tau = m/r$ is the time constant of an oscillator of mass m and damping constant r (the damping force per unit velocity). The time constant determines how quickly the oscillator comes into a steady state when it is driven from rest by a harmonic driving force and how quickly the stored energy drains away once the driving force is removed. At resonance, the power is related to the driving force F by

$$P = \frac{F^2\tau}{2m}.$$

In general, the energy stored in an oscillator as a function of the driving frequency follows a Lorentzian curve centred on the resonant frequency and having width $\Delta\omega/\omega = 1/\omega\tau$.

13.1   The opera singer and the wine glass

It is claimed that an opera singer can break a wine glass by singing a loud sustained note. In this question we investige the physics behind this feat.

(a) What causes a brittle material such as glass to break? (See tutorial 26.)

(b) Explain why the claim specifies a wine glass rather than any other sort of glass.

(c) Why does the note have to be sustained?

(d) What determines which note should be sung and how important is it to hit the right note accurately?

(e) Suppose that an opera singer can produce sound with intensity $L_p \approx 100$ dB at the glass and that the glass will break if it is dropped on to the floor from a height of about 0.5m or more. Estimate (i) the power in the sound produced by the singer and (ii) the amount of sound energy from the singer that can be stored in the glass. Hence decide whether the glass will break.

Note that L_p is related to air pressure by $L_p = 20 \log_{10}(p/p_0)$ where the reference pressure $p_0 = 2 \times 10^{-5} \, \text{N m}^{-2}$.

13.2   The walls of Jericho

According to the biblical story (Joshua 5-13) the walls of Jericho were brought down by a host of trumpeters blowing their trumpets for seven days. Excavation has revealed that the walls were built of mud bricks. By physical reasoning decide whether this is a believable story or whether it is more likely that the trumpeters were brought down.

13.3   Keep on swinging

Children quickly learn how to build up and maintain the motion of a playground swing. They pull on the supporting chains of the swing at the bottom of its travel and relax the pull at the extremeties of the motion. This pull raises the swing slightly at the bottom of its arc and the work done feeds energy into the swinging motion. The technique is known as parametric amplification. It is also used to maintain the swing of the Foucault pendulum at the Science Museum in London (*Interdisciplinary Science Reviews*, 1994, **19**, 326). The object of this question is to investigate how the method works.

(a) Consider a simple pendulum of length l and bob mass m. Ignoring any damping, show that the tension in the wire is given by
$$T = mg \cos \theta + ml\dot{\theta}^2$$
and hence that, for small angles of swing θ with $\theta = \theta_0$ at $t = 0$, the tension can be written
$$T = mg \left[1 + \frac{1}{4}\theta_0^2 - \frac{3}{4}\theta_0^2 \cos 2\omega t \right].$$
where $\omega = \sqrt{g/l}$.

Now let the suspension head of the wire be supported by a mechanism which gives a vertical displacement $-a \sin 2\omega t$.

(b) Obtain the mean power delivered to the pendulum by the mechanism.

(c) Show that the energy stored in the pendulum increases exponentially in the absence of damping.

13.4   Coupled oscillators

(a) Two oscillators with the same natural frequency ω are coupled in such a way that their displacements $x_1(t)$, $x_2(t)$ satisfy

$$\ddot{x}_1 + w^2 x_1 = gx_2$$
$$\ddot{x}_2 + w^2 x_2 = gx_1$$

where $g \ll w^2$ (so the coupling is weak). One is set in motion by giving it a small displacement such that $x_1(0) = 2a$, $\dot{x}_1(0) = 0$, while the other is initially undisplaced, $x_2(0) = 0$, and at rest, $\dot{x}_2(0) = 0$. Show that the subsequent motion is given by

$$x_1 = a(\cos w_1 t + \cos w_2 t)$$
$$x_2 = a(\cos w_1 t - \cos w_2 t)$$

where

$$w_1 = \sqrt{w^2 - g} \simeq w - \frac{g}{2w},$$
$$w_2 = \sqrt{w^2 + g} \simeq w + \frac{g}{2w}.$$

Hence describe how the motion of each oscillator varies with time.

(b) The following diagram shows a set of pendulums of various lengths and an additional single pendulum at C hung from a common bar. If pendulum C is set swinging, describe the subsequent motion of the pendulums.

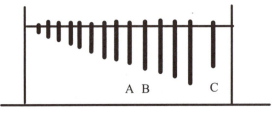

A B   C

Figure 4. The pendulum at C has a length intermediate between that of A and B

(c) In order to describe the interaction of an atom with the electromagnetic field, the electronic structure of the atom can be modelled as an electric dipole oscillator with a moment $d(t) = ex\,(t)$ that varies sinusoidally in time, i.e. the atom (in this model) is a simple harmonic oscillator (figure 5). This is weakly coupled to the electromagnetic field, which can be described as an infinite set of harmonic oscillators having all possible frequencies. If this system were treated according to classical mechanics what would be the fate of an isolated atomic oscillator according to the results of (a) and (b)?

(d) In quantum theory any oscillator in its ground state has a zero point energy. This applies to both the atomic oscillator and the field oscillators, even if the classically measured value of the field is zero. Explain why, in this quantum picture, the ground state of an isolated atom is stable.

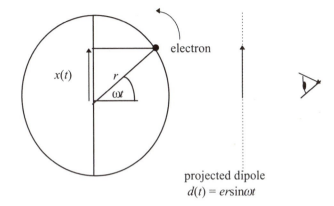

*Figure 5. An electron in an atomic orbit can be pictured as an oscillating di-
pole. A multi-electron atom would be represented by the net dipole resulting
from the contribution of each electron. The atom can therefore be represented by
an atomic oscillator.*

Tutorial 14   Antigravity

14.1   Cavorite

In H. G. Wells's novel *The First Men in the Moon* professor Cavor invents a material - cavorite - which can shield against gravity just as a conductor can shield
from electrostatic fields. Electrostatic shielding is possible because there are
both positive and negative charges in nature. So gravitational shielding requires
the existence of matter possessing negative mass. The object of this question is
to show that the existence of particles of negative mass would not only enable
us to shield against gravity but also to extract limitless amounts of energy from
the interaction of the two sorts of mass.

(a) Show that, according to Newtonian dynamics, two particles of negative mass
will repel each other.

(b) There is no problem with this result. It is in the interaction of a negative mass
particle with a positive mass particle that the problem arises. Show that the two
particles would both accelerate in the same direction. What laws of mechanics
would such behaviour violate?

In relativistic quantum field theory the existence of negative mass would allow the extraction of unlimited energy from the vacuum (which is, in principle, an infinite resource) without violating any laws. Whether this is either true in nature or practicable is not clear.

14.2   Cosmological constancy

Although discrete particles having negative mass appear to be ruled out by the above considerations, a uniform distribution of negative mass is not. A cosmological constant is equivalent to such a uniform distribution of negative mass. In the Newtonian limit of general relativity with a cosmological constant the force on a planet of mass m in orbit about the Sun of mass M is

$$F = -GmM/r^2 + \Lambda mr/3,$$

where Λ is the cosmological constant and r the distance between the bodies.

(a) Show that for Λ positive the cosmological term is equivalent to a distribution of negative mass of density $\rho = -\Lambda/4\pi G$.

(b) Cosmological observations limit the value of Λ to be not greater than $\approx 10^{-35}\text{s}^{-2}$. By how much would such a cosmological constant affect the period of the Earth's orbit about the Sun?

14.3   Bubble gravity

It might be thought that spherical bubbles of vacuum in an infinite self-gravitating sea of uniform mass density (like Dirac's holes) would constitute a model for the behaviour of negative masses. Show this is not the case by considering the motion of two such bubbles under their mutual gravitational interaction. (Neglect such complications as the mutual deformation of the bubble shapes.)

Tutorial 15   Orbits of the Sun

Strictly speaking Kepler's laws of planetary motion apply to the motion of an isolated body round the Sun. But Kepler showed they hold to a good approximation for individual bodies in the many-body solar system. This tutorial shows how the use of Kepler's laws, together with conservation of energy and angular momentum, simplifies the solution of many orbit problems. For reference, Kepler's laws are (i) Planets move in ellipses with the Sun as a focus (ii) The area swept out per unit time is constant (iii) $T \propto a^{3/2}$ (where T is the period of the orbit and a the semi-major axis of the ellipse.)

15.1 The wasting Sun

The Sun radiates energy and hence mass, since mass and energy are equivalent. It also loses mass through the solar wind at the rate of about 2×10^{-14} solar masses year^{-1}.

(a) Which of these two processes dominates?

(b) The loss of mass by the Sun will change the radius of the Earth's orbit. Show, however, that $MR = $ constant, where M is the mass of the Sun and R the distance of the Earth from the Sun.

(c) Hence decide whether changes in the Earth's orbit over geological time scales could give rise to changes in the Earth's climate, assuming that the present solar mass loss rate applies in the past.

15.2 Orbital impact

There are strong grounds for believing that 65 million years ago an asteroid of about 12 km in diameter collided with the Earth and gave rise to widespread extinctions through its immediate effects on the Earth's climate. Could the asteroid have created a small permanent change in the Earth's climate by altering the Earth's orbit? To answer this question suppose that the asteroid were travelling at the same speed as the Earth but in the opposite direction and so made a head on collision with the Earth. Take the Earth's initial orbit to be a circle and take the density of the asteroid to be the same as that of the Earth. Find the Earth's new orbit and hence answer the question.

15.3 Hale-Bopp spectacular

Comet Hale-Bopp, the most spectacular comet of the second half of the twentieth century, was 0.914 AU from the Sun at the perihelion (closest point) of its orbit on 2 April 1997 at which point its velocity relative to the Sun was 44.01 km s^{-1}.

(a) What is the escape velocity at perihelion?

(b) Determine the comet's distance from the Sun at aphelion (i.e. at the furthest point of the orbit) Hint: use the appropriate conservation laws to express the aphelion distance as a function of the perihelion distance and of the ratio of the escape velocity at perihelion to the actual velocity at perihelion. If you stay alert this task can be accomplished without the need to solve a quadratic equation!

(c) Estimate the year in which it next passes through perihelion.

(d) Why is your estimate likely to be innacurate?

This explains why the orbits of long period comets cannot be determined even approximately from naked eye observations and hence why Halley's comet was so important to the development of Newtonian gravity (*Scientific American*, September 1998).

Tutorial 16   Accelerated frames

Newton's laws, strictly speaking, describe motion relative to inertial frames of reference only. Yet we commonly ignore the fact that the surface of the rotating Earth is not an inertial frame and use Newton's laws with impunity. Here, and in the following tutorial, we shall investigate effects arising from the acceleration of the frame of reference and see when we can, and when we cannot, ignore them.

16.1   Nevertheless it moves

A medieval argument against the rotation of the Earth was that a stone thrown up into the air does not land to the west of its launching point. This argument uses Aristotelian physics, according to which horizontal motion requires a force to maintain it and as soon as the force is removed the body comes instantaneously to rest. But the vertical motion is supposed to be maintained for a time, to be consistent with observation.

(a) How far to the west might a stone thrown vertically to a height of 10 m be expected to land according to Aristotelian physics? (For the sake of this argument assume the vertical motion is under constant downward acceleration.)

(b) How far to the east or west would the stone land under Newtonian physics?

The answer is not exactly zero because the Earth's surface accelerates relative to the stone whilst the stone is in the air. We shall call this the *ground acceleration effect*.

16.2   The latitude effect

Another cause of deviation due to the rotation of the Earth is the latitude effect, which is generally much larger than the ground acceleration effect evaluated in question 1(b). The latitude effect arises because the velocity of a point on the Earth's surface depends on its latitude.

(a) Suppose that a stationary battleship in the northern hemisphere fires a shell in a direction due north at a stationary target located due north. Even if the range is exactly right the shell will miss its target. To see this, view the Earth from a frame of reference in which the only motion of the Earth is its rotation on its axis, and assume this frame is, to sufficient accuracy, an inertial frame. If the latitude of the battleship is 45^0 estimate the amount by which the shell misses its target if the speed of the shell is 300 m s^{-1} and the distance of the target is 18 km.

(b) Repeat this calculation for a golf ball struck a distance of 250 m the time of flight of which is 6 s.

So viewed from the Earth the shells from naval guns and other projectiles behave as if there is a force acting on them causing them to veer from a straight line, although they are travelling on a straight line relative to an inertial frame. Such an apparent force arising from the non-inertial nature of the reference frame is called a coriolis force. Naval gunners allow for it by aiming off the target and golfers ignore it.

16.3   The first battle of the Falklands

In the first world war battle of the Falklands the British fleet had had to rush from the northern to the southern hemisphere to engage the German fleet. On opening the engagement the British gunners observed their shells landing very wide of their targets despite careful aiming. Explain what had gone wrong.

16.4   Accelerations of the Earth

It is not quite accurate to take the only non-inertial motion of the Earth to be its rotation. What other accelerations (relative to an inertial frame) does the Earth have?

Tutorial 17   Artificial Gravity

An example of a non-inertial frame which is simpler than one attached to the surface of the rotating Earth is one fixed to the inner surface of the giant space-ship described in Arthur C. Clarkes novel *Rendezvous with Rama*. Rama is a hollow cylinder of radius 10 km and length 50 km which rotates with a period of 4 minutes. The spin provides an artificial gravity for the passengers who live on the inner curved surface of the cylinder. There is no latitude effect on a rotating cylinder as all points on its curved surface have the same velocity about its axis. But the other effects arising from the acceleration of the reference frame are present.

17.1   Recreational sports on Rama

(a) Show that the inhabitants of Rama who live on its inner curved surface experience only a negligible true gravitational force arising from its mass. Assume the material of Rama has a density of 8500 kg m^{-3}.

(b) What is the effective gravitational acceleration at the inner surface of Rama arising from its rotation?

(c) A projectile is launched at time $t = 0$ from the inner surface of Rama with a velocity $v_\perp$ normal to the 'ground' which is small compared with its rotational velocity. Show that the height y of the projectile at time t is given approximately by

$$y = v_\perp t - \tfrac{1}{2}gt^2$$

where g is the effective gravitational acceleration on Rama. (Thus, the rotation of Rama mimics gravity for the particular case of the motion of a projectile.)

(d) Suppose we decide to pass our time on Rama by playing golf. Then the effect which causes objects thrown up on the Earth's surface to be displaced on landing will come into play (the ground acceleration effect of tutorial 16). To see the importance of this effect in a game of golf calculate the deviation suffered by a drive struck down the centre of a fairway aligned parallel to the cylinder axis. Take the initial speed of the ball to be 50 m s^{-1} and the angle of launch as 25° and ignore lift forces on the ball.

(e) What would be the effect of the rotation of Rama on a long putt hit along a line parallel to the axis of the cylinder?

(f) If we find golf too difficult under these conditions we can always take up tennis where the time of flight (typically less than 1s) of the ball is much shorter than in golf. However, another effect of the accelerating frame has to be reckoned with: When serving the ball is launched from a height about 2 m above ground level which is 2 m closer to the cylinder's axis. By how much will this cause a serve to deviate for a court oriented along the length of the cylinder?

Tutorial 18   Stress and strain

Solids subject to tension or to compression generally obey Hooke's law. This states that the applied force per unit area (the stress) is proportional to the resultant fractional change in length (the strain). The constant of proportionality is the elastic modulus which is characteristic of the material in question. Hooke's law arises from the form of the forces between atoms which are to a good approximation harmonic for small displacements (see tutorial 12 on harmonic motion). Thus Hooke's law is an approximation which is valid up to strains of about 1%;

beyond this the stress-strain relationship is no longer linear. Within the Hookean regime solids exhibit elasticity: that is, the deformation they suffer is reversible so they recover their original form when the applied stress is removed. The work that is done in deforming a solid is stored as strain energy in the Hookean regime and is mostly recoverable. When subjected to strains greater than about 1% the work done produces irreversible changes in the solid, and at even greater strains they break (see tutorial 4).

18.1   Towing ropes

(a) Show that the elastic energy E stored in a stretched Hookean rope is given by

$$E = \frac{1}{2}Y \times (\text{strain})^2 \times \text{volume of rope.}$$

(b) Hence explain why a long towing rope is less likely to be damaged or to break than a short one.

(c) In the film Airport-Concorde, a Concorde aircraft with its braking system and reverse thrusters put out of action by a terrorist attack is stopped at Orly airport by a sequence of elastic landing nets deployed, one after the other, along the runway. Estimate how much rubber would be required to do this, and hence decide whether this is a feasible tactic for stopping runaway aircraft. Rubber can store a great deal of strain energy before it breaks, say 4×10^7 J m^{-3}. The mass of Concorde is in the region of 100 tonnes and its landing speed around 240 km hr^{-1}.

18.2   Jumping flea

From slow motion film it is known that a jumping flea accelerates from rest to a velocity of 1m s^{-1} in 10^{-3} s. The mass of a flea is 0.45×10^{-6} kg and the maximum power from insect muscle is 60 W kg^{-1}.

(a) Compare the acceleration of the flea with that of a human Olympic high jumper. The Olympic high jump record is 2.4 m.

(b) How much kinetic energy does the flea acquire in a jump?

(c) If 20% of the flea's body weight is muscle can muscle alone power the jump?

(d) At the base of each hind leg of the flea is a pad of volume 1.4×10^{-4} mm^3 of a substance, resilin, having a Young's modulus of 1.7×10^6 N m^{-2}. Show that enough energy can be stored by compressing the resilin pads to power the jump.

(e) The resilin pads must be compressed by muscular effort, so how is the problem in part (c) circumvented?

18.3   Into the deep

The bottom of the Marianas Trench in the Pacific Ocean is 10.9 km below sea level.

(a) Estimate the pressure at the bottom of the trench. The density of sea water at the surface is 1025 kg m^{-3}.

(b) Would taking account of the change in density of sea water with depth make a significant difference to the pressure obtained in the above estimate? The bulk modulus of sea water is 2.1×10^9 Pa.

(c) A bathyscaph is a spherical diving vessel designed to descend to great depths in the ocean. Estimate the thickness of steel wall needed to withstand the pressure at the bottom of the Marianas Trench. Take the internal radius of the bathyscaph to be 1 m and the bulk modulus for steel to be 2×10^{11} N m^{-2}. Hint: consider the force pushing one hemisphere of the vessel into the other hemisphere.

This descent was carried out in 1960 by Jaques Picarde.

Tutorial 19   Hot and cold

Heat flows through a body of area A and thermal conductivity k in which there is a temperature gradient $\Delta T/\Delta x$ at the rate $kA(\Delta T/\Delta x)$ W. A perfect black body at temperature T radiates at a rate σT^4 W m^{-2}.

19.1   Physics of dieting

(a) Estimate your daily energy requirement assuming this is needed only to supply heat lost by radiation to your surroundings.

(b) Hence, roughly how much weight could you lose in a week by fasting? (Fat has an effective energy content of 24×10^6 J kg^{-1}.)

19.2   Saving the *Titanic* ?

The survival of icebergs beyond the polar regions makes them a hazard to shipping. Various methods to get rid of them have been proposed. One method is to cover the iceberg with a thick layer of soot delivered from an aircraft. The blackened iceberg will then absorb all the sunlight incident on it and melt rapidly. Assuming that this strategy could be implemented decide how effective it would be. Take the mass of the iceberg to be 100,000 tonnes and make reasonable assumptions about its shape. The latent heat of fusion of water is 333 kJ kg^{-1} and the solar constant is 1360 W m^{-2}.

This experiment has been tried by the International Ice Patrol. The results were inconclusive (see Encyclopaedia Britannica).

19.3   Winter sports

(a) Why does a pond freeze from the top down?

(b) What thickness of ice could form in one cold night with the air temperature at -10C? The thermal conductivity of ice is about 2 W K^{-1}m^{-1}.

(c) A layer of ice 30cm thick on a pond is (probably!) safe to skate on. If this is the case, how many days of freezing weather is needed to make ice skating safe?

19.4   Warming the Earth

(a) The average temperature at the surface of the Earth is 30 °C lower than that at a depth of 1 km. Given that the thermal conductivity of the Earth's crust is 2 W K^{-1} m^{-1} find the rate of heat loss per m^2 from the Earth.

(b) The Earth's crust contains small amounts of the long lived radioactive elements uranium, thorium and potassium the decay of which produces heat, on average, at the rate 4×10^{-10} J s^{-1} per kilogram of rock (*Physics of Planetary Interiors*, G H A Cole, 1984). Show that the concentration of these radioactive elements in the Earth's interior cannot be as high as it is in the rocks of the crust. The density of the Earth is 5500 kg m^{-3}

The reason for this segregation of the radioactive elements is not known (see the above reference).

Tutorial 20   Cooling the Chunnel

It is possible to convert mechanical work into heat with virtually 100% efficiency, as, for example, when work is done against a frictional force. There is, however, an upper limit to the efficiency of an engine that turns heat energy into directed mechanical work. This upper limit is achieved by an ideal heat engine operating in a Carnot cycle between a hot source and a cold sink. Bearing in mind that the performance of a real heat engine will fall short of that of an ideal engine we shall use the theory of the Carnot engine to investigate the problem of cooling the Channel rail tunnel that links Dover to Calais.

A great deal of electrical energy is used in the Channel tunnel system nearly all of which degrades to heat and has to be removed. In each of the two running tunnels cooling water is circulated through a pipe which extends to the half way point and then returns back to the tunnel entrance. The heat transfered to the water is extracted in a chiller plant (at the base of Shakespeare Cliff on the British side) after which the chilled water is returned to the tunnel. A similar system operates on the French side cooling the other half of each tunnel. The object is to find the theoretical minimum power needed to run the chiller plant.

You will need the following data. It takes 25 minutes for a train to pass through a tunnel and trains follow each other at 3 minute intervals. The mean power driving a train is 4MW. You can ignore other sources of heat input and heat loss from the tunnels. Chilled water enters the tunnels at a temperature of 5 °C and leaves at a steady temperature of 30 °C. The specific heat of water C is 4190 J kg^{-1} K^{-1}.

(a) What is the rate of flow of water in kg s^{-1} through the cooling plant at the base of Shakespeare Cliff?

Consider an ideal refrigerator, which is an ideal engine run backwards. Let it cool M kg of water from an ambient temperature T_0 to a temperature T_f, ΔT below ambient temperature.

(b) Show that the energy E supplied to the refrigerator is given by
$$E = MCT_0 \log_e (T_0/T_f) - MC\Delta T.$$

(c) Hence show that for $\Delta T \ll T_0$
$$E \approx \frac{1}{2} MC\Delta T^2/T_0.$$

(d) The first stage of the chiller plant is a heat exchanger which brings the water to the ambient temperature before it passes into the refrigeration plant. For an ambient temperature of 20 °C use the results obtained above to find the power needed to operate an ideal refrigeration plant under these conditions.

Tutorial 21   Heat flow in ice-houses

Before the invention of the refrigerator ice gathered from frozen ponds in winter was stored in ice-houses. All large country houses had such an ice-house which provided them with a supply of ice during the rest of the year. A typical ice-house (see the diagram) was a large brick cavity-walled chamber partially or fully sunk into the ground and well insulated. The ice was broken up before loading so that it formed a compact mass inside the chamber. At the base of the chamber there was a drain hole to take away the melt water which would otherwise accumulate in the bottom of chamber and spoil the heat insulation provided by the straw bundles which lined the chamber's inner surface.

(a) The rate at which the ice in an ice-house melts, in units of mass per unit time $\dot{M}$, in terms of the rate at which heat flows into the ice $\dot{E}$, the rate at which water evaporates from the ice $\dot{M}_V$, the latent heat of evaporation L_V and the latent of melting L_M is given by *the ice-house equation*

$$\dot{M} = \frac{\dot{E} - \dot{M}_V L_V}{L_M}.$$

Use the consevation of mass and energy to derive this equation.

(b) *The importance of evaporation.* From the equation derived in part (a) it is clear that there are two extremes of ice-house design. There is the air tight ice-house in which no evaporation to the outside takes place and, at the other extreme, never realised in practice, is the ice house where all the melted ice evaporates. How much faster is the rate of melting in the first case than it is in the second case, all other things being equal? The latent heat of melting is 0.33×10^6 J kg^{-1} and the latent heat of evaporation is 2.26×10^6 J kg^{-1} .

(c) Ice-houses came in many shapes and sizes so we will consider a conveniently shaped cylindrical storage chamber of diameter 3.5 m and depth of ice, when full, of 5 m. Assuming that heat is conducted into the ice through the cavity wall and straw lining with which it is in contact, and that no ice is taken out of the ice-house, integrate the ice-house equation to get the mass of ice in the chamber as a function of time. [Hint: account for evaporation by assuming that a fraction f of melted ice evaporates.]

(d) Estimate the half-life (the time for half the ice to disappear) of this ice-house from the following information. The measured temperature at the surface of the ice is about 3 °C. (*The Ice-Houses of Britain*, S. P. Beamon and S. Roaf, 1990) Take the heat conduction rate through the cavity wall and straw lining

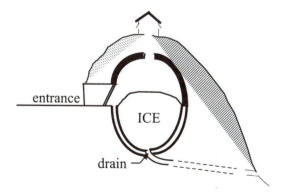

entrance

ICE

drain

Figure 6. The structure of an ice-house

to be $1 \text{ W m}^{-2} \text{ K}^{-1}$ which is the value appropriate to an insulated cavity wall, the insulation in this case being provided by the inner lining of straw. Take the density of ice in the chamber to be 800 kg m^{-3} and the temperature of the ground to be 8 °C, the typical temperature of an underground cellar.

Tutorial 22   The atmosphere

The variation of pressure P with height z above the ground in an isothermal atmosphere in hydrostatic equilibrium is given by
$$P = P_0 \exp(-mgz/kT) \qquad (1)$$
where m is the mean molecular mass of the atmosphere, T its temperature and the atmosphere satisfies the perfect gas equation of state $P = \rho kT/m$. This is required in questions 2 and 3.

22.1   Columns of air

(a) Given that the pressure of the atmosphere at sea level is 1.013×10^5 Pa obtain, the mass of the column of air extending from the top of your head to the limit of the Earth's atmosphere without making any assumptions about the temperaure distribution.

(b) Why are you not crushed by the weight of this air?

(c) If the Earth's atmosphere were to be liquified how deep an ocean covering the whole Earth would it form? The density of liquid nitrogen is 800 kg m^{-3}.

(d) Explain why there is an upper limit to the pressure that an atmosphere can exert at the surface of a planet.

22.2   Climbing Everest

The ability of the human body to absorb oxygen depends on the atmospheric pressure. The body can acclimatise to altitude but physiological limit of adaptation to low pressure appears to be reached if the pressure drops to 0.33×10^5 Pa. (Science, **187**, 313). The object of this question is to decide whether it is possible for anyone to survive, for more than a brief spell, at the summit of Mt. Everest (height 8848m) without breathing apparatus.

(a) Obtain an answer to the question by using equation (1). The sea level temperature of the US Standard Atmosphere is $T_0 = 288$ K.

(b) If your life depended on it you would not trust this answer as equation (1) applies strictly to an isothermal atmosphere, whereas from sea level to a height of 11 km the temperature of the Earth's atmosphere falls linearly at a rate of

6.50 K km^{-1}. Show that, when this temperature gradient is taken into account, the pressure as a function of height for an atmosphere of mean molecular mass m is given by

$$P = P_0(1 - \alpha z/T_0)^\lambda, \qquad (2)$$

where $\alpha = dT/dz$ and $\lambda = mg/k\alpha$.[Hint: solve the equation of hydrostatic balance with $T = T_0 - \alpha z$.]

(c) Evaluate the pressure at the summit of Everest using equation (2). This is a more accurate result than that obtained using the simpler isothermal atmosphere and enables you to answer the original question.

Tutorial 23   Atmospheric friction

The Chicxulub crater situated near the Yucatan penninsula in Mexico was produced by the meteoritic impact of an asteroid estimated to be about 12 km in diameter (*Nature*, 1997, **390**, 472). This event occured 65 million years ago and may have been the cause of the mass extinctions of animal and plant species that occured at about the same time. The impact is conjectured to have thrown up dust and sulphur compounds into the stratosphere which would have cut down the amount of solar radiation reaching the ground for several years.

(a) Estimate the lowest and highest speed that such an object could have just prior to its entry into the Earth's atmosphere.

At such hypersonic speeds atmospheric friction heats up a meteorite to the point where it loses mass through ablation (the entrainment of melted material) from its surface. As a result of this process of mass loss the atmosphere shields us against the hoards of low mass objects which impinge on the Earth every day by burning them up before they reach the ground.

(b) Explain why more massive meteorites do in fact survive to reach the ground.

(c) Write down expressions for the mass lost by a meteorite through ablation. [Hint: consider the encounter in the rest frame of the meteorite and estimate the energy that can be transferred to it from the air stream.] Hence estimate the mass loss for a 12 km meteorite travelling at 20 km s^{-1} incident normally on the atmosphere and show that it is a negligible fraction of the intial mass.

Take the density of the meteorite to be 2500 kg m^{-3}, the heat of ablation to be 5×10^6 J kg^{-1} and the heat transfer coefficient (the fraction of energy disipated that is absorbed by the meteorite) to be 0.02.

(d) Show that the atmospheric drag force on a body of area A and velocity v moving through air of density ρ is proportional to $\rho A v^2$.

The constant of proportionality is usually written as $\frac{1}{2} \times C_d$, where C_d, the drag coefficient, has a value in the range 0.5 to 2.0.

(e) Hence show that a body of 12 km diameter would not be slowed down appreciably by the atmosphere.

(f) Estimate the maximum aerodynamic pressure on the body and decide whether it would break up before impact with the ground. The crushing strength of recovered meteorites ranges up to 500 MPa.

(g) Compare the kinetic energy per unit mass of a meteorite travelling at 20 km s^{-1} at impact with the ground with the energy yield per kg of high explosive. (See tutorial 1).

(h) How does the crater size depend on the kinetic energy at impact?

(i) Why are there no impact craters on the ocean floor (*Nature*, 1997, **390**, 340)?

Tutorial 24   Bernoulli's theorem

Bernoulli's theorem is a statement of the conservation of energy for the steady motion of a fluid. It states that at any point along a stream line of a fluid the pressure P, density ρ, velocity u, and height h obey the relationship

$$P + \rho g h + \rho u^2/2 = \text{constant.}$$

This tutorial will illustrate some applications of Bernoulli's theorem.

24.1   Mir

On 25 June 1997 the Russian space station Mir was holed in a collision with a supply vessel. The internal atmospheric pressure of Mir fell from its normal value of 750 mm mercury to 675 mm mercury in about 8 minutes. Given that the pressurised volume of the space station was 390 m^3 and the internal temperature of Mir was 24 °C, estimate the size of the hole.

24.2   Aerodynamic lift

In flight an aircraft wing changes the direction of the air flowing around it giving this air a downwards component of velocity as shown in the figure.

(a) Explain how this flow gives rise to a lift force on the wing.

(b) Show that the lift is proportional to the (area of the wing) × (density of the air) × (velocity of the wing through the air)^2.

This redirection of the airflow by a wing lowers the pressure above the wing relative to the pressure below it.

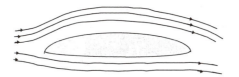

Figure 7. Streamlines over a thin wing

(c) Estimate this average pressure difference for the case of a Boeing 747 which has a fully loaded weight of 330 tonnes and a wing area of about 500 m^2.

(d) By how much is the air flowing across the top surface of the wing speeded up relative to the air flowing under the wing?

24.3   The flight of a golf ball

(a) A fact well known to golfers is that a properly struck golf ball spins as it flies through the air. This spin gives rise to a lift force on the ball that is at right angles to its spin axis and to the direction of motion. Note that the surface of the ball must be roughened to achieve the effect, hence the dimpling on a golf ball. Sketch the air flow pattern round a spinning golf ball in flight and relate the direction of the resultant force to the direction of rotation.

This force is known as the Magnus force after H.G. Magnus, a German scientist who investigated the effect in the mid nineteenth century.

(b) In a well hit golf drive the initial velocity and rate of spin are 50 m s^{-1} and 30 rev s^{-1} respectively. Estimate the corresponding lift force in terms of the weight of the ball. The mass and radius of a golf ball are 0.046 kg and 2.1×10^{-2} m and the density of air is 1.2 kg m^{-3}. [Hint: Take the difference in air flow velocity across the ball to be equal to the difference in the rotational velocity of opposite sides of the ball and treat the golf ball as a cylinder spinning about its axis.]

Tutorial 25   Spontaneous structure

Mechanical systems relax to a state of lowest energy, U. Thermodynamical systems in equilibrium adopt the configuration of lowest *free energy*, $F = U - TS$. Under the right conditions this can lead to the formation of spontaneous structures.

25.1   Ordering life?

(a) As a preliminary we need a few facts about free energy. The free energy per

particle of a perfect gas of density n at temperature T is
$$f_{gas} = kT(\log_e(nv_0) - 1),$$
where $v_0 = h^3/(2\pi mkT)^{3/2}$. If the gas molecules have internal energy ε show that the free energy per particle becomes
$$f = kT(\log_e(nv) - 1) + \varepsilon.$$

(b) Consider now a solvent medium with a concentration of surfactant molecules. These are molecules with polar and non-polar parts which, in a solvent medium, have a tendency to associate into mutually non-interacting clusters. The clusters can therefore be considered to behave as particles of a perfect gas. Let a cluster of N surfactant particles have internal energy ε_N. To favour aggregation ε_N must decrease with N, at least for small N. i.e. the energy cost of producing larger aggregates decreases with the size of the aggregate. For simplicity let the volume occupied by each surfactant molecule equal that occupied by each solvent molecule. Let the number of surfactant molecules in a cluster of size N be $P_N \times$ (the total number of surfactant and solvent molecules). Show that the free energy per particle is
$$F = \sum_N \frac{P_N}{N} \left\{ kT \left[\log_e \left(a \frac{P_N}{N} \right) - 1 \right] + \varepsilon_N \right\}$$
where a is a constant.

(c) To find the aggregation state adopted by the system we minimise the free energy per particle F with respect to the P_N at constant T, subject to $\sum P_N = $ constant. Show that this gives
$$P_N = aNe^{N(\mu - \varepsilon_N/NkT)},$$
where $\mu = $ constant is the Lagrange multiplier for the constraint.

Note that the system does not clump into two spatial separate phases (aggregate and solvent) because there is an entropy (order) cost in so doing. If the free gas entropy term were omitted, this is precisely what would happen.

(d) Suppose now that only two values of N are possible, so molecules are either unaggregated or form structures of M molecules. Show that at low concentrations (μ is small) the system is mostly mono-molecular, whereas at high concentrations molecules aggregate. (*Statistical Thermodynamics of Surfaces, Interfaces and Membranes*, S. Safran, 1994, Addison-Wesley).

This is a relatively simple example of how systems can spontaneously order. This particular example may be of significance in the spontaneous formation of pre-cellular life if abiotically produced long-chain molecules aggregate to form 'proto-cells'.

25.2   Oil on troubled water

Waves are produced on the surface of water by the action of wind if the wind speed is high enough. In fact, if the water density is ρ, the density of the air ρ' and the wind velocity is U, the surface is unstable to rippling if

$$\rho' U^2 > \rho V^2$$

where V is the speed of surface waves. (This says that the momentum flux in the wind must exceeds that in the waves it is producing if these are to exist.) For waves of wavelength λ, V is given by

$$V^2 = \frac{g\lambda}{2\pi} + \frac{2\pi T}{\rho\lambda}$$

where T is the surface tension and g the acceleration due to gravity.

(a) Taking $\rho = 10^3$ kg m^{-3}, $\rho' = 1.26\times10^{-3}\rho$, and $T = 7.4 \times 10^{-2}$ N m^{-1}, show that according to this theory, waves are created if $U \gtrsim 6.5$ m s^{-1}(Kelvin *Collected Papers*, 1910, vol iv p76)

For the shortest wavelengths the surface tension dominates gravity. Such waves are called ripples. In fact, ripples occur for lower wind velocities than in Kelvin's theory, which Kelvin attributed to the viscosity of air. (See *Hydrodynamic and Hydromagnetic Stability*, S. Chandrasekhar, 1961, Oxford, for more details.)

(b) If oil is poured on to water in a mono-molecular layer rippling is suppressed for a given wind velocity. The surface tension of oil is around 3×10^{-2} N m^{-1}. Why does this appear to raise a problem for the theory?

(c) Experiments show that the surface tension of a mono-layer decreases significantly as it is stretched and increases as it is compressed. How, qualitatively, could this fact be used to resolve the problem?

Tutorial 26   Material strength

26.1   All cracked up

(a). Given that the heat of formation of a crystal of silver from the dispersed ions is 3.0 eV per molecule estimate the breaking stress of a silver wire. (Silver crystallises in a FCC structure with a nearest neighbour distance of 2.88Å.)

The strength obtained in this way is some 30 000 times the experimental value of 2.9×10^8 N m^{-2}. This is because the strength of real materials is determined by the density of microscopic cracks which act as stress raisers. This information can be used to estimate the length of a typical crack as in the following question.

(b) Let a glass plate of unit thickness having Young's modulus of elasticity

$Y = 7 \times 10^{10}$ N m^{-2} be subject to an extensive stress σ_0 (N m^{-2}). Show that the presence of a thin crack across the thickness of the glass of length l orthogonal to the direction of stress in the plate reduces the strain energy by an amount $l^2\sigma_0^2/Y$.

(c) The crack can propagate *without external work* if the increase in surface energy for an incremental increase δl in the length of a crack is provided by a reduction in the strain energy of the plate. Show that this leads to

$$\delta \left(\frac{l^2\sigma_0^2}{Y} \right) = 2T\delta l$$

where T is the surface tension of glass.

(d) For glass $T = 0.56$ N m^{-1} and $\sigma_0 = 7 \times 10^7$ N m^{-2}. Obtain the typical length of cracks in glass.

The original theory is due to A A Griffith, (*Trans Roy. Soc.*, 1920, **A221**, 163,). By drawing thin fibres of glass cracks orthogonal to the fibre axis can be eliminated and the tensile strength greatly increased.

26.2   Building in brick

Bricks have a density of 2000 kg m^{-3} and a maximum working compressive strength of 10^6 N m^{-2}. What is the tallest building that can be constructed entirely of brick?

26.3   Up in smoke

According to a certain TV documentary crashing a jet fighter at Mach 1 into an immovable wall releases enough energy to completely vapourise the jet. Is this correct?

26.4   Flowing glass

Glass has an amorphous structure, rather than the crystalline structure we usually associate with a solid, and can be thought of as a supercooled liquid. Thus, given sufficient time, glass should flow. It is often claimed that evidence for this can be seen in old window glass which is sometimes thicker at the bottom than at the top. It is difficult to measure the viscosity of glass at room temperature because it is so high, but above a few hundred degrees C the viscosity η, as a function of temperature, of GeO$_2$ glass is given by

$$\log \eta = -9.94 + \frac{1792}{T}$$

where T is measured in degrees C. By working out the temperature at which the flow of window glass would take a few hundred years decide whether the flow of glass accounts for the shape of old window panes.

Tutorial 27   Electric shocks

27.1   Electric sight

In the late 1870s Edison's invention of the electric lightbulb caused shares in gas companies to plummet. This prompted William Preece, who was later to become engineer in chief of the British General Post Office, to publish some calculations which, he claimed, proved that centrally supplied electricity could never replace gas as a source of domestic lighting. His conclusion was that the energy dissipated per lamp would be inversely proportional to n^2, where n is the number of lamps (*Nature*, 1897, **19**, 261-262. See also *Heaviside, Sage in Solitude*, P Nahin.). The object of this first question is to see why Preece's conclusion was wrong for the small supply utilities that were envisaged at the time.

Consider the simple d-c circuit which was considered by Preece and is shown in the figure. The voltage V drives current through n light bulbs each of resistance R which are wired in parallel; r is the resistance of the wires and r_i the internal resistance of the power source.

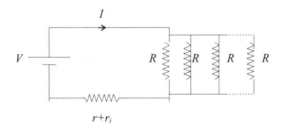

Figure 8. An electricity supply circuit

(a) Derive an expression for the power dissipated in each bulb.

(b) Calculate the resistance of a 60 W domestic light bulb.

(c) In Edison's original electricity supply utility $r + r_i$ was a fraction of an ohm and the resistance of an Edison lamp was about 200 Ω when hot. Use these values, together with the answer to (b), to sketch the energy dissipated in a lamp as a function of n, the number of lamps, and hence estimate how many light bulbs would have to be in the circuit to make Preece's conclusion valid.

(d) How do modern power suppliers avoid a large power drop to the consumer at times of high demand?

27.2   Working at high potential

Repairs to the high voltage overhead power lines of the national grid can be carried out without having to switch off the power. The linesman who works on the live power lines is clad in an electrically conducting suit and works from a metal trolley which is hooked on to the live line. The trolley and its occupant are lowered on to the line on the end of a non-conducting rope by a helicopter. Once the trolley is hooked on to the line sparks start to jump between the line and the linesman's hands and elbows. The sparking ceases when the suit is electrically connected to the line by a conducting lead.

(a) What causes the sparking and why do the sparks jump between the sharp parts of the linesmans anatomy and the high voltage line?

(b) Estimate the electrical capacity of the isolated linesman. This obviously depends on his/her shape so assume that this is spherical.

(c) Hence estimate the current which flows through the cable connecting the linesman to the high voltage power line. Take the line votage to be 400 kV and the frequency to be 50 Hz.

(d) Explain why the linesman when clad in his/her suit is safe from electrocution.

27.3   Electric defibrillation

Heart defibrillators, which are used to restore a regular heart beat, stimulate the heart to contract by delivering a short current pulse of duration $\approx$20 ms. In one type of defibrillator a capacitor is charged to a suitable voltage and then discharged through the patient's chest with the aid of two large electrodes. The defibrillator needs to be able to deliver pulses of up to 360 J to patients with transchest resistances ranging up to 150 Ω. Estimate values for capacity and voltage needed to cope with these requirements.

Tutorial 28   Magnetic fields

28.1   Lightning strikes

The current in a lightning conductor following a strike can be sufficient to cause a hollow cylindrical conductor to crumple. This information can be used to obtain an estimate, albeit crude, of the current in a lightning flash.

(a) Suppose the lightning conductor consists of a long thin cylindrical conducting shell of outer radius a that carries a uniform current I. What is the order of magnitude of the magnetic field in the shell?

(b) What is the pressure exerted by the magnetic field on the shell ?

(c) From the fact that it is possible to crush a thin copper tube by stamping on it, estimate the pressure required and hence obtain the current.

28.2   Magnetic pliers

Sometimes when lightening strikes the conductor of a building it has been ob-served that nails can be pulled out of the floorboards. A possible explanation of this occurrence is that the magnetic field of the lightning current magnetises the nails and is then responsible for extracting them. A *uniform* magnetic field can-not exert a net translational force on a nail; it can only exert a torque, like that on a compass needle by the Earth's field. However the gradient of a *non-uniform* field can give rise to a translational force. The question arises as to whether this explanation is plausible.

(a) Show that a magnetic field $\mathbf{H}$ acting on a magnetic dipole of moment $\mathbf{m}$ gives rise to a force $-\mathbf{m} \cdot \nabla \mathbf{H}$.

(b) Show that the magnetisation $\mathbf{M} = \mu \, \mathbf{H}$ for iron ($\mu = 1000$) would exceed the saturation magnetisation of 1 Bohr magneton per atom.

(c) Decide whether this force is sufficient to explain the observation.

Tutorial 29   Model circuits

We use a simple LCR circuit (an inductor, capacitor and resistance in series) to illustrate what can go wrong when the limitations of a mathematical model are ignored.

29.1   The CR circuit

(a). A CR series circuit is connected to a battery, voltage V_b by closing a switch. Once the connection is made what is the energy delivered by the battery in charg-ing the capacitor? Compare this with the energy stored in the capacitor when it is fully charged to the voltage V_b and show that dissipation in the resistor during charging accounts for the missing energy.

(b) Now let the resistance in the circuit tend to zero. There is still the same missing energy but, since the resistance is now zero, the circuit appears to be non-dissipative! How is this contradiction resolved? [Hint: begin by considering the charging timescale.]

29.2   The LCR circuit

(a) Now consider an open LCR series circuit, in which the capacitor is initially uncharged and the current is obviously zero, which can be connected to a battery by closing a switch. Show that if $R^2C > 4L$ the current in the circuit at time t after the circuit is made is

$$I = \frac{\lambda_2 CV_b}{(\lambda_1 - \lambda_2)}e^{\lambda_1 t} - \frac{\lambda_1 CV_b}{(\lambda_1 - \lambda_2)}e^{\lambda_2 t},$$

where

$$\lambda_1 = -\frac{R}{2L} - \sqrt{\frac{R^2}{4L^2} - \frac{1}{CL}},$$

$$\lambda_2 = -\frac{R}{2L} + \sqrt{\frac{R^2}{4L^2} - \frac{1}{CL}}.$$

(b) Show that as $t \to 0$ we have $I \to 0$ for any small but finite value of L.

(c) Compare this result with the initial current in a CR circuit, where L is exactly zero. How is it that an infinitesimal change in the physical properties of a circuit, which causes L to acquire a non-zero value, no matter how small, can make a finite difference to the current?

A practical consequence of this is that the solution for small L cannot be obtained by expanding about the solution for $L = 0$. A more familiar example is the turbulent flow of fluids. An ideal fluid (one with exactly zero viscosity) can never be turbulent, whereas a real fluid, no matter how small its viscosity, will always exhibit turbulence for a sufficiently large Reynolds number.

Tutorial 30   Dispersion on the line

Electromagnetic waves travel through a vacuum with a speed c that is independent of their wavelength. Thus (from Fourier theory) a wave of arbitrary shape will propagate without distortion. Such dispersion free propagation is unusual. In general, the components of a wave group travel at different speeds and the group spreads out as it propagates. For example, waves travelling through a medium or down a cable can suffer dispersion. Here we consider dispersion in the context of transmission lines.

30.1   Thomson's speechless cable

One of the greatest engineering feats of the 19th century was the laying of submarine cables under the Atlantic ocean. The first such cable was layed in 1858
and had a length of about 3600 km. The theoretical analysis of signal propagation along a cable was first carried out by William Thomson (later Lord Kelvin).
He derived the following equation for the voltage $v(x,t)$ on the line

$$\frac{\partial^2 v}{\partial x^2} = RC\frac{\partial v}{\partial t},$$

where R and C are respectively the resistance and capacitance per unit length of
the line. Note that Thomson regarded the inductance of the line as unimportant
and ignored it. This equation has the wave solution

$$v(x,t) = \text{const.} \times \exp\{-(\omega RC/2)^{1/2}x\}\cos\{\omega t - (\omega RC/2)^{1/2}x\}.$$

(a) Deduce that these waves are dispersive.

(b) Show that, according to Thomson's theory, the transmission of intelligible
speech across the Atlantic via a cable would not be possible. Note that speech
requires frequencies in the range 1 to 4 kHz. Typical parameters for an early
submarine cable are $C = 7.5 \times 10^{-11}\text{F m}^{-1}$ and $R = 7 \times 10^{-3}\Omega\text{ m}^{-1}$. [Hint:
Compare the propagation times of two frequencies within the speech range.]

(c) Show that the transmission of intelligible morse code is possible according
to the theory.

(d) Show that Thomson's equation allows waves of arbitrarily high velocity to
propagate down the line.

This is an unphysical feature of the diffusion equation. The next question shows
how this defect in the theory was remedied.

30.2   Heaviside — speaking

In 1876 Oliver Heaviside generalised Thomson's equation (question 1) by including the inductance L of the line. His equation is

$$\frac{\partial^2 v}{\partial x^2} = RC\frac{\partial v}{\partial t} + LC\frac{\partial^2 v}{\partial t^2},$$

which has the solution

$$v(x,t) = v_0 e^{-\gamma x}\cos(\omega t - kx),$$

where $\gamma^2 = (RC\omega/2)\{(1 + L^2\omega^2/R^2)^{1/2} - L\omega/R\}$ and $k = RC\omega/2\gamma$.

(a) Show that in the limit that $L\omega/R \ll 1$, the solution of Heaviside's equation
reduces to that of Thomson's equation.

(b) The inductance of the early submarine cables was typically

$L = 4.6 \times 10^{-7} \, \mathrm{H\,m}^{-1}$. Show that Thomson's theory describes the transmission of Morse code signals correctly.

(c) On the other hand, Heaviside deduced that, if $L\omega/R \gg 1$, then to a good approximation there is no dispersion of signals on the line and the attenuation of signals is independent of frequency. Show this.

To achieve this condition Heaviside proposed that inductors be distributed along the cable. When this was adopted transmission of speech down telephone cables became possible.

(d) For a coaxial line, show that Heaviside's equation does not permit the speed of waves on the line to exceed the speed of light. For a coaxial line of inner and outer radii a and b, $C = 2\pi\varepsilon\varepsilon_0/\log_e(b/a)$ and $L = (\mu\mu_0/2\pi)\log_e(b/a)$.

Tutorial 31   Things to do with mirrors

31.1   Solar furnaces

A parabolic mirror brings parallel light incident along its axis to a single focus. The Sun appears to us as a disc of apparent angular diameter $\alpha = 0.5^0$. Thus every element of surface of a parabolic mirror illuminated by the Sun acts as a source of a diverging cone of rays. A parabolic mirror will therefore produce an image of the Sun of finite size at its focus. If the atmosphere is even slightly dusty this image will appear as a small sphere of light hovering in space at the focus.

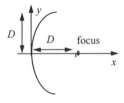

Figure 9. A parabolic mirror with aperture equal to its focal length

(a) Estimate the diameter of this nearly spherical image for the mirror shown in the figure.

(b) Show that the solar constant can be written $\frac{1}{4}\alpha^2\sigma T_s^4$, where T_s is the surface temperature of the Sun.

(c) Hence, estimate the energy per second incident at the image and show that the energy per unit area averaged over the image surface is independent of the distance of the mirror from the Sun.

(d) Estimate the maximum temperature to which a blackbody placed at the focus of the mirror (assumed perfect) could be heated.

(e) This temperature is close to the surface temperature of the Sun. Could a lens or system of lenses and mirrors produce a temperature above that of the Sun, or would such an outcome be in conflict with the second law of thermodynamics? The assumptions of ray optics will cease to be valid at some point and the wave nature of light must be taken into account.

(f) At what distance from the Sun does the conclusion of part (c) break down for a mirror of diameter 3m?

31.2 Going to war with mirrors

Archimedes is supposed to have set fire to the Roman fleet at the Siege of Syracuse by directing sunlight on to the ships by means of mirrors.

(a) Assuming that the ships come within 50 m of the shore, and assuming that Archimedes had a perfect parabolic mirror of focal length 50m, what size would it need to be in order to set fire to a ship? (The flux needed to ignite wood within 10 s is 6.3×10^4 W m^{-2}.)

(b) Of course, Archimedes would not have had such a mirror. Legend has it that he used the flat burnished shields of the defending soldiers which would be non-focussing and imperfect reflectors. Is this legend credible?

(Reflections on the 'Burning Mirrors of Archimedes', A.A. Mills & R Clift, *Eur.J.Phys.*, 13, 268-279, 1992).

Tutorial 32   Blackbody Radiation

All bodies above absolute zero radiate. When this radiation is able to come into thermal equilibrium with the bodies emitting it, the radiation has a characteristic spectrum, the Planck spectrum, which depends only on the temperature T of the system. Such radiation is called blackbody. It has an energy density aT^4 [J m^{-3}]. The energy per unit time per unit area emitted by a body in thermal equilibrium with the radiation is σT^4 [W m^{-2}]. The average energy of a blackbody photon is $2.7\,kT$. On the long wavelength (low frequency) side of the peak of the Planck distribution the blackbody spectral intensity can be approximated by the Rayleigh-Jeans law

$$I_v = \frac{2v^2}{c^2} kT \text{ [W m}^{-2}\text{ster}^{-1}\text{Hz}^{-1}].$$

32.1   'Blackish' bodies

(a) Why is neither a gas flame nor the Sun a perfect blackbody? Which of them is the closer approximation?

(b) Why can you not listen to the gas flame on your radio?

(c) Radio telescopes detect radio emission from distant quasars having angular sizes less than 1mas ($=10^{-3}$ seconds of arc) with a spectral flux of 10^{-29} W m^{-2}Hz^{-1} at GHz radio frequencies. By estimating the temperature of the blackbody required, show that the radio emission is unlikely to come from material in thermal equilibrium.

32.2   Photon numbers

(a) Estimate the number density of blackbody photons in a blacked-out room at 20 °C and compare with the number density of air molecules in the room.

(b) The result of (a) gives the impression that matter is a more important constituent of the Universe than radiation. Show this is false by comparing the number density of photons in the cosmic background radiation (which is an almost perfect blackbody at 2.7 K) with the mean number density of atoms. The mean universal mass density of matter is about 10^{-26} kg m^{-3}.

(c) In some circumstances the presence of thermal photons can play a significant role. For example, in the LEP accelerator at CERN high energy electrons and positrons circulate around a circular beam pipe which is highly evacuated to minimise collisions with air molecules. Repeat the calculations of part (a) for the interior of the LEP beam pipe. Take the temperature to be 20 °C again and the pressure inside the beam pipe to be 10^{-10} Torr (1 Torr = 1 mm of mercury).

(d) What effect will the blackbody photons have on the circulating electrons and positrons?

32.3   The Sun as a blackbody

(a) Given that the Sun is yellow in colour and assuming that it radiates as a blackbody, obtain a value for the radius of the Sun. The Solar constant is 1360 W m^{-2} and Wien's displacement law is $\lambda_{max}T = 2.9 \times 10^{-3}$ K m.

(b) Could this method be used to estimate the radius of the red star Betelgeuse (the top left star in the constellation of Orion)?

32.4 The Jupiter star

In Arthur C. Clarke's novel 2010 the planet Jupiter is turned into a star. (It is not made clear how!)

(a) Estimate the maximum energy flux that would be incident on the Earth from Jupiter under these circumstances. For the sake of argument take the surface temperature of Jupiter to be the same as that of the Sun. The radius of Jupiter's orbit is 5.5 a.u. and the mean radius of Jupiter is 6970 km.

(b) Suggest ways in which life on Earth would be affected by the Jupiter star.

Tutorial 33   Relativistic travel

In this tutorial we illustrate some of the consequences of time dilation, length contraction and the relativity of simultaneity.

(a) Show that, in principle, there is no upper limit to how far a human can travel despite the fact that humans are mortal and speeds greater than that of light are not possible.

(b) Suppose that in 500 years time it becomes possible to build a spaceship capable of travelling the 2×10^{22} m distance from the Earth to the Andromeda galaxy in 50 years of spaceship time. Ignoring the time to accelerate and decelerate the spaceship, what would its relativistic γ-factor be? Compare this with the γ-factor of protons in the 8 TeV Large Hadron Collider at CERN.

(c) What would be the earliest date on which news of the safe arrival of an expedition setting out in 2500 would be received back on Earth?

Thus we see that, although intergalactic travel is possible in principle, the time dilation that accompanies space travel would make it an impractical venture for humans to undertake.

(d) What is the distance between Earth and Andromeda in the rest frame of the astronauts once they have reached full speed ?

Tutorial 34   Hyperbolic motion

A body moving with constant *proper acceleration* in a straight line is said to be executing hyperbolic motion because its position and time are related by the equation of a hyperbola

$$x^2 - c^2 t^2 = c^4/a^2,$$

where a is the proper acceleration of the body and x and t are its position and time as measured with respect to its initial rest frame S. Hyperbolic motion has some interesting features which we are going to investigate in this tutorial. The tutorial will also illustrate the use of spacetime diagrams.

34.1   Constant acceleration in relativity

(a) Draw the world line of a body moving with constant proper acceleration on a spacetime diagram labelled with the spatial coordinate x and the time coordinate t of the inertial frame S. Draw in the asymptote of this world line and decide what it represents physically.

(b) Add the axes of an instantaneous rest frame S of the body subsequent to the start of the acceleration. Show that, in its instantaneous rest frame, the distance of the body from the origin event remains constant as long as the acceleration is maintained.

34.2   Siamese rockets

Consider now a well known example that tests our understanding of the difference between Newtonian and relativistic kinematics. Two identical rockets are joined by a taut and inelastic cable as shown in figure 10.

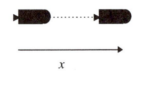

x

Figure 10. Siamese Rockets

The motors of both rockets are started at the same instant in their initial rest frame and they accelerate identically in the x-direction until their fuel is used up. By the end of these manoeuvres the cable has broken. Take the cable to be much longer than the rockets so that any changes to the lengths of the rockets can be ignored.

(a) Explain this outcome from the point of view of the initial rest frame of the rockets. [Hint: Draw the world lines of the rockets on a spacetime diagram.]

(b) At what point in the journey does the cable break ?

34.3   Rigid bodies

Suppose instead that both rockets execute hyperbolic motion tending towards a common asymptote as shown in the spacetime diagram (figure 52).

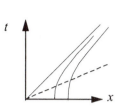

Figure 11. Hyperbolic motion

(a) Show that the rear rocket is accelerating faster than the front rocket.

(b) Show that the distance between the two rockets in their instantaneous rest frame does not increase, so a cable connecting the rockets would not break.

(c) Show that for a given acceleration of the front rocket, there is a maximum separation that the two rockets can have, which occurs when the rear rocket is forced to travel at the speed of light.

Tutorial 35   Mechanics near c

35.1   The size of particle accelerators

The proposed Large Hadron Collider, which is to be built at the European Centre for Nuclear Research CERN near Geneva will accelerate protons to a maximum energy of about 8 GeV.

(a) Given that the protons in the LHC are constrained to move along a circular path by 9 T bending magnets, use relativistic mechanics to calculate the radius of the path.[Hint: use the fact that $\gamma \gg 1$.]

(b) Now repeat the calculation but assume that Newtonian mechanics applies.

The calculation shows that such a Newtonian machine would be a good deal cheaper to build, but we know from experience that it would not work. The great size of particle accelerators is a dramatic illustration of the inadequacy of Newtonian mechanics for speeds approaching that of light.

35.2   Relativistic snooker

A useful rule of thumb for snooker players is that the angle between the directions taken by two snooker balls after a glancing collision is 90 degrees (see tutorial 3). It is interesting to investigate whether the same rule can be used in the relativistic version of the game. To answer this question, consider the particular collision shown in figure 18 where both balls go off at an angle θ with the initial direction of the cue ball.

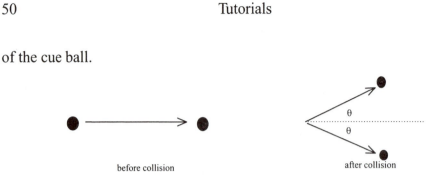

Figure 12. Relativistic snooker balls

35.3   A relative paradox

A body, viewed from its rest frame S', emits some of its internal energy in the form of isotropic radiation and conserves momentum by remaining at rest in S'. The body is now observed from a frame S moving with respect to S'. The radiation is no longer emitted isotropically, but has a higher intensity in the direction of motion of the particle. The radiation therefore has a net momentum in this frame. It would therefore appear that the body must slow down to conserve momentum in the frame S. But if it does so, it cannot remain at rest in S' as we have deduced. We have therefore arrived at a contradiction. Find the error in the reasoning and show that momentum is conserved in both frames.

Relativity abounds in such apparent paradoxes which usually arise from making a Newtonian assumption somewhere in the argument.

Tutorial 36   A relativistic aberration

In their novel *Into Deepest Space* Fred and Geoffrey Hoyle describe the experiences of astronauts in a spaceship which accelerates to near light speed away from our Galaxy. Before high velocities have been reached the travellers have a wonderful face on view of the galaxy through their rear window. By the time they are approaching the speed of light the Galaxy has disappeared from their rear window and is seen strangely altered through the front window. This is an example of aberration which is the change in direction of a moving object or light ray that accompanies a change of reference frame. (A more familiar, non-relativistic instance of this phenomenon is the change in direction of falling snow when viewed from a moving car rather than from the ground.) In the first problem we investigate how, by travelling fast enough, we can see something ahead of us even though we are moving away from it.

The diagram shows a spaceship stationary above the plane of the galaxy. A light ray coming from the edge of the visible disc of the galaxy is shown intercepting the spaceship at an angle θ with respect to the x axis. Now imagine we jump into a moving spaceship at the same location.

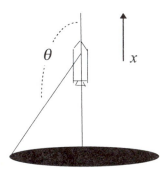

Figure 13. A spaceship accelerating away from the Galaxy

(a) Use the relativistic velocity transformation to find the angle θ' that the light ray makes with the x'-axis of the moving spaceship frame, where the x-axes of the two frames are parallel to each other.

(b) How fast would the spaceship have to travel for an angle $\theta = 150^0$ to be rotated into an angle $\theta' = 30^0$?

This explains how the effect described in *Into Deepest Space* comes about.

(c) Show that in order to aberrate a ray having $\theta > 90^0$ into a ray with $\theta' < 90^0$ the spaceship must overtake the ray in question.

(d)Why cannot the centre of the Galaxy be seen ahead?

(e) What would the Galaxy look like when seen from the high speed spacecraft ?

(f) Describe how the appearance of the Galaxy would change as the distance of the space craft from it is increased without any change in its speed?

(g) If the spaceship accelerates too slowly it will not reach a speed to aberrate sufficiently an ever receding galaxy into the forward direction. What acceleration would be required to obtain the effect?

Tutorial 37   Lensing gravity

For this tutorial we need one result from the general theory of relativity which deals with the effects of gravity on light. The theory predicts that when a ray of light passes a spherically symmetric body of mass M it suffers an angular

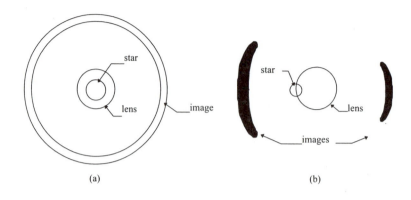

(a)   (b)

Figure 14. (a) At perfect alignment the lensed image is an Einstein ring. (b) Out of alignment image

deflection θ towards the body, which for small deflections, is given by

$$\theta = \frac{4GM}{c^2 r}, \tag{3}$$

where r is the distance of closest approach to the centre of the body. It follows from this that, if a massive dark object passes between an observer and a distant star, light from the star can be bent round the lensing object into the eye of the observer. If, at some instant, the star, lens and observer are perfectly aligned, the image of the star at this conjunction will be a thin circular ring, known as an Einstein ring, centred on the lens as shown in figure 14(a). (When alignment is not perfect two images of the star are formed on either side of the lens as shown in figure 14(b).)

Lensing events have been detected (*Nature*, 1993, **365**, 623) in which an object, probably a brown dwarf or a faint low mass star, in the halo of our galaxy has passed near to the line of sight of a star in the Large Magellanic Cloud (LMC).

(a) The distance to the LMC is about 50 kpc. Assuming that a lensing mass is one fifth of the way from the solar system to an LMC star calculate the diameter of the Einstein ring that would be formed at perfect alignment.

(b) The Hubble space telescope has a 3 m diameter mirror. Decide whether it could resolve the Einstein ring formed in the above event.

(c) Would you expect the colour of the ring to differ from that of the lensed star?

Equation (3) shows that the deflection of a ray decreases with increasing r. Despite this, rays impinging on the lens plane at different values of r can still be brought to a focus at the observer if the angle which the rays make with the normal to the lens plane decreases with increasing r. The star, which is the source of the light being lensed, has a finite size. Therefore rays coming from different points of its surface will impinge on the lens plane at different angles, so a ring of finite angular thickness will be formed, as shown in figure 15.

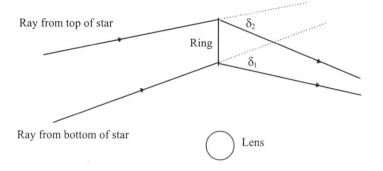

Figure 15. The lens produces convergence of these rays even though the outer ray is refracted less than the inner one.

(d) Use the figure to show that the angular thickness of the Einstein ring at the observer is equal to the angle subtended by the radius of the unlensed star at the observer.

(e) Why are the brightnesses of the star and the ring the same?

(f) By what factor will lensing increase the intensity of light received from the star by the observer when the star is aligned behind the lens?

Tutorial 38   The Universe

38.1   Olber's paradox

If the universe were uniformly populated with stars stretching to infinity in all directions, as casual observation suggests, then every line of sight from the Earth would end on the surface of a star and the sky would be uniformly ablaze with light. The fact that the sky is dark at night tells us that there is something wrong with this apparently reasonable deduction. This conflict between theory and observation is known as Olber's Paradox although in fact it was first stated by Kepler. Lord Kelvin argued in 1901 that the night is dark because the stars have a finite age. The *visible* universe has a finite size, which is determined by the speed of light and the age of the stars, and therefore only a fraction of the sky is covered by stars (*Phil. Mag Ser* 6, 1901, **2**,161-167). The aim of this question is to reconstruct Kelvin's argument, using modern values for the ages of the stars, and to show that it resolves the paradox.

(a) Ignoring the expansion of the universe find the fractional coverage of the sky by stars assuming them to be spread out in space uniformly, as Kelvin did, rather than clumped into galaxies. Take the age of the stars to be 10^{10} years, the mean density of matter in the universe to be 2×10^{-26} kg m^{-3} and the mass and diameter of a star to be 2×10^{30} kg and 7×10^8 m respectively.

(b) How far out would the visible universe need to stretch for a full coverage of the night sky by stars, and what age would the stars have to have?

(c) An alternative way of looking at the problem is to realise that, if the sky were as bright as the surface of a star, the universe would be filled with radiation at the same temperature as the stars. This approach should give the same age as was obtained by the method in question (b). To demonstrate this consider a star occupying a volume equal to the average volume per star. Assume that the volume is surrounded by reflecting walls and estimate the time it takes for the star to fill the volume with radiation to its own temperature.

(d) What effect would the clumping of stars into galaxies have on the above arguments?

(e) And what effect does the expansion of the universe have on the arguments?

Note that expansion is not the main cause, and therefore it is not required in order to resolve the paradox in a universe which has a positive deceleration. In other cases it is important e.g. in the steady-state universe - see Harrison (*Nature*, 1991, **352**).

38.2   Faster than light?

A more modern paradox involves the speed of recession of objects in an expanding Universe. Uniform expansion implies that the speeds of distant galaxies can exceed the speed of light, a result that appears to be in conflict with the theory of relativity. In fact, there is no conflict as this question aims to show.

(a) Imagine a currant loaf of unlimited size which expands uniformly as it cooks. Every currant sees every other currant receding from it. Show that the velocity of recession v of any currant from a chosen currant is proportional to their distance d apart i.e. show that $v \propto d$.

(b) Now, by replacing the currants by galaxies and the dough by space, we make the transition to the Einstein - De Sitter expanding universe. This model of the universe has Euclidean spatial geometry, and expands in the same uniform manner as the currant loaf. It is currently the theoretically favoured description of the actual universe. The relation between the proper distance d of a galaxy at any time and its recession velocity v at that time is therefore

$$v = Hd$$

where H is the Hubble parameter at that time (H is a function of time) The present distance in an Einstein- De Sitter universe to a galaxy having an observed redshift z is

$$d = (2c/H)[1 - (1 + z)^{-1/2}].$$

Beyond what value of z is the recession speed greater than c?

(c) What is the recession speed of the remotest object we can in principle (just) see?

The previous two questions show that recession speeds in cosmology (defined as in (b)) can be greater than c. This would be inconsistent if, as assumed in special relativity, inertial frames of unlimited spatial and temporal extent could be set up.

(d) Explain why such a global inertial frame cannot be constructed in an expanding universe.

(e) How does the result of (d) resolve the apparent conflict with special relativity?

38.3   Gamma-Ray bursts

Bursts of γ-rays from space are detected at the rate of about one per day coming from random directions in the sky. Assuming that these bursts are produced in galaxies out to redshift $z = 2$, estimate how long we shall have to wait for such a burst to occur in our own Galaxy. The average density of galaxies at the present time is about 0.03 Mpc^{-3}. Assume an Einstein-De Sitter universe and take H_0, the current value of the Hubble parameter, to be 60 km s^{-1}Mpc^{-1}.

Tutorial 39   The anthropic principle

The idea behind the anthropic principle is that the features of the universe that are necessary to support life can, in the end, be traced to the values of the fundamental physical constants. The fact that we are here then 'explains' why these constants have the measured values. This tutorial shows how this works in the case of the lifetime of the Sun, as far as is possible without involving detailed stellar structure theory. Our basic assumption is that the development of higher forms of life requires the lifetime of the Sun to be at least 10^9 years.

(a) The Sun is characterised by its mass M, and its radius, R. It is composed of a perfect gas of mainly ionised hydrogen. Radiation is scattered by electrons and therefore prevented from escaping out directly from the interior. This keeps the centre hotter than the surface, such that the black body energy density at the surface is of order aT^4/τ, where T is the central temperature and τ is the optical depth to the centre. (The optical depth is the radius R expressed as a number of mean free paths for the radiation; a is the radiation constant). For simplicity assume that radiation is scattered only by isolated free electrons for which the scattering cross section is πr_e^2, where r_e is the classical electron radius.

(i) Show that the average density is of order M/R^3

(ii) Assuming the Sun is in hydrostatic equilibrium, show that the central pressure is of order $P \sim GM^2/R^4$.

(iii) Show that the central temperature is given by $kT \sim Gm_pM/R$.

(iv) By considering the flow of energy through the surface show that, to order of magnitude, the luminosity is $L \sim (G^4 m_p^5 m_e^2/\alpha^2 h^5)M^3$, where $\alpha = e^2 h/(4\pi\varepsilon_0 c)$ is the fine structure constant.

(b) This is as far as one can get without an estimate of the typical mass of a star. This can be obtained by requiring that the gas pressure in a star exceeds the radiation pressure (since otherwise the system is unstable). Show that this requires $M \lesssim (hc/G)^{3/2}m_p^{-2} \sim M_\odot$ (the mass of the Sun).

(c). Hence show that the lifetime of the Sun in terms of fundamental constants is $t \sim Mc^2/L \sim e^4/[(4\pi\varepsilon_0)^2 Gc^3 m_e^2 m_p] \sim 10^{10}$ years.

(d) If the Sun were composed of deuterium the nuclear energy would be generated via the strong interaction (^{2}H, ^{2}H) → (^{3}H, p) instead of the weak reaction (p,p)→(^{2}H,e,ν). What effect would this have on the lifetime of the Sun?

Note that the result in (c) depends only on fundamental constants and is indeed quite sensitive to some of them: a factor 10 change in e/m for the electron or in the speed of light would make life impossible, since no star would shine for long enough to enable higher forms of life to develop. This is surprising: in order for us to be here the Universe has to be finely tuned! A priori, apparently a universe with life is highly unlikely. The anthropic principle turns this round: because we find we are here we must also find appropriate values for the fundamental constants. Thus, in a sense, the anthropic principle explains these values.

Tutorial 40   Radiating gravity

The inverse square law for force as a function of distance is formulated in terms of distances measured at some absolute time. Such a law cannot stand on its own in a relativisitic theory. In electromagnetic theory the electrostatic force follows an inverse square law, but the theory as a whole is compatible with relativity because moving charges are subject to an additional magnetic component of the force. The electric and magnetic components are not separately Lorentz invariant, but the combined force is, and therefore does not depend on an absolute time. Mathematically, the scalar electrostatic potential appears as the fourth component of the 4-vector potential which generates both E and B fields. This suggests for Newtonian gravity to be compatible with relativity there must exist extra components to the gravitational force. A first guess might be that, like electromagnetism, the true gravitational potential should be a 4-vector. The binary pulsar can be used to show this is not the case.

The binary pulsar consists of two neutron stars in orbit, one a pulsar which provides an accurate timing mechanism. The system radiates energy in the form of gravitational disturbances propagating at the speed of light (gravitational waves) and the consequent loss of gravitational potential energy from the binary leads to a measurable decay of the orbit. The object of the first problem is to compute the radiation from the binary pulsar using an analogue of the Larmor formula from electromagnetic theory to describe the radiation. This would correspond to a theory of gravity with a vector potential. The resulting orbital decay is inconsistent with observation, ruling out such a theory.

40.1   Dipole radiation

(a) In electromagnetic theory the force between like charges e is an inverse square law $\propto e^2/(4\pi\varepsilon_0 r^2)$. The Larmor formula for the luminosity of electromagnetic radiation from a charge e with acceleration $\ddot{x}$ is $L = \frac{2}{3}e^2\ddot{x}^2/(4\pi\varepsilon_0 c^3)$. In the Newtonian theory of gravity the force between like masses M is an inverse square law $\propto GM^2/r^2$. What would be the analogous formula for the gravitational luminosity of an accelerated mass?

(b) Let the binary pulsar be approximated by two equal point masses in circular orbits about a common centre with a separation equal to the major axis of the true orbit. Show that the gravitational radiation in part (a) would give rise to a decay in the period T of the orbit of $dT/dt = -8.0 \times 10^{-9}$ s s^{-1} (about 3000 times the observed value).

The following data is given for the binary pulsar (more, and to a higher accuracy, than is needed for the question). The pulsar mass is 1.42 $M_\odot$, the companion mass is 1.40 $M_\odot$, the inclination i of the orbit (the angle between line of sight and normal to the plane of the orbit) is given by $\sin i = 0.76$, the semi-major axis a of the pulsar orbit is given by $a\sin i = 2.34185$ light-seconds, the period is 27906.98163 s, the orbital eccentricity is 0.617127 and the shift in the periastron of the pulsar orbit is 4.2263 minutes of arc per year. The binary pulsar is at a distance of about 5 kpc.

40.2   Quadrupole radiation

(a) The general theory of relativity describes the gravitational field by a second rank tensor (the 10 components of the metric). In this theory the power in gravitational radiation L_{grav} turns out to depend on the moment of inertia tensor I of the radiating system (quadrupole radiation). In fact $L_{grav} \propto \overset{...}{I}{}^2$. From this information, without going into details, it is possible to construct an order of magnitude estimate for the gravitational radiation power as an exercise in dimensional analysis. Check that your result is about 6 orders of magnitude less than the formula you derived in (1).

(b) The observed rate of change of orbital period is -2.4×10^{-12} s s^{-1}. What would be the expected energy flux at the Earth if the decay of the orbit is due entirely to gravitational radiation?

Tutorial 41   Radioactive decay

The radioactive decay law, that the number of decays per second is proportional to the number of particles present was discovered by Rutherford and Soddy in the course of their investigations into the decay of radioactive nuclei. The law, a consequence of quantum mechanics, also applies to the de-excitation of excited atoms and to the decay of unstable fundamental particles. The law is statistical in nature, that is, it does not predict when a given particle will decay but only gives the probability of its decay in a given time interval. For a collection of particles it predicts, subject to statistical fluctations, the number of decays in a given time interval.

41.1   A natural fission reactor

The Oklo uranium mines in West Africa are unusual in that some of the uranium ore found there is heavily depleted in the ^{235}U isotope. The world average abundance of ^{235}U is 0.72% whereas at Oklo it is typically 0.4%. A study of the site has led to the conclusion that, at some time in the remote past, the presence of water in the uranium deposits created conditions similar to those in a man-made, water-moderated nuclear reactor, and that criticality lasted long enough to cause the depletion of the ^{235}U isotope. Nuclear reactors moderated with ordinary water are fueled with uranium in which the proportion of ^{235}U has been enriched to 3-5%. This is nessessary because a chain reaction in present day natural uranium is not possible in the presence of water on account of the comparitively high neutron capture cross section of the hydrogen protons in the water. Estimate how long ago it is since natural reactors such as the one at Oklo existed.

The half life of $^{238}U = 4.5 \times 10^9$ years, that of $^{235}U = 7.13 \times 10^8$ years

41.2 The Uranium clock

Today natural Uranium consists of 99.28% ^{238}U and 0.72% ^{235}U. Assuming that nuclear reactions in a single supernovae created these isotopes in equal proportions, how long ago were they made?

Note that the Earth's uranium was probably not all created in a single supernova event so this calculation gives the mean age. The assumption of equal yields for the two isotopes is a reasonable approximation. (D. Schramm & E. Symbalisty, *Rep Prog. Phys.*, 1981).

41.3 Natural Pu

Plutonium is usually refered to as a man made element but this is not strictly true. Estimate an upper limit to the amount of ^{239}Pu in equilibrium with 1 kg of natural uranium.

The spontaneous fission half-life of ^{238}U is 10^{16} yr and about 2.5 neutrons are produced on average per fission.

Tutorial 42 Elementary particles

42.1 Catching neutrinos

In the Homestake mine neutrino detector on average 0.5 atoms of ^{37}Ar are created per day through the capture of neutrinos from the Sun in ^{37}Cl. At intervals the accumulated radioactive ^{37}Ar atoms are flushed out of the detector and counted.

(a) How many argon atoms are present in the detector when radioactive equilibrium has been established?

(b) How many argon atoms would accumulate in the detector over a period of 3 months - the usual interval between flushings?

The half life of $^{37}Ar = 37$ days.

The observed rate of creation of ^{37}Ar atoms is about one third of the rate predicted for the standard solar model. This discrepancy is known as the solar neutrino problem.

42.2   The long-lived proton

(a) Some current theories of fundamental particles predict that protons are not stable, but decay into other particles at a very slow rate. It is therefore interesting to see if we can put limits on the decay rate of the proton. If the decay of protons in our bodies were to give us a dose of more than about 50 mSv in a year, the maximum yearly exposure permitted by US government regulations for workers exposed to radiation, then we should 'know in our bones' that proton decay was affecting our health. Use the value of this dose to estimate a lower bound to the lifetime of the proton. Assume that the relative biological effectiveness (RBE) factor of the proton decay products is 1 and take the body to be composed of water.

A dose of 1 Sv equals an energy of $1 \, \mathrm{J \, kg^{-1}} \times$ RBE.

(b) Grand unified theories of elementary particles predict proton lifetimes which are very much longer than the lifetime estimated in part (a), so proton decay detectors have to be much larger than the human body. They consist of a large volume of water sited deep underground surrounded by detectors which can monitor the Cerenkov radiation emitted by the fast positron coming from a postulated decay reaction $p \rightarrow \pi^0 + e^+$. What mass of water would be needed to detect one decay per year if the proton life time were 10^{33} years?

Tutorial 43   Quantum uncertainty

As a consequence of the Heisenberg uncertainty principle the more closely an electron is confined to a region of space the higher its kinetic energy will be. In an atom the electrons are confined by the Coulomb potential of the nucleus. The competition between the confining nature of the potential and the liberating tendency of the uncertainty principle gives rise to various quantum mechanical effects.

43.1   An alternative Bohr

(a) Use the uncertainty principle to estimate the kinetic energy of an electron confined within a given radius in a hydrogen atom.

(b) Hence estimate the size of the hydrogen atom in its ground state by minimizing its total energy as a function of the orbital radius of the electron.

(c) Compare the size and ground state energy obtained in this way with the values obtained from a Bohr theory calculation (M. Bowler, *Femtophysics*, Pergamon, 1990).

For one electron systems the Bohr theory gives the same energy levels as the Schrödinger equation.

43.2   Pressure Ionisation

When atoms are subjected to a high enough pressure they become ionized. This will happen, for example, at the centre of a sufficiently massive gravitating body.

(a) In order to ionize an atom a certain minimum energy must be supplied to it, 13.6 eV in the case of hydrogen. Estimate the reduction in atomic radius required to ionize a hydrogen atom.

(b) What pressure P is needed to bring this about? (Hint: $P = -\partial E/\partial V$ where E is energy and V volume.)

(c) A planet is defined as a body in which the *atoms* resist the compressive force of gravity. Estimate the maximum mass and size of a planet composed of hydrogen. [Hint: You will need to *estimate* the pressure required at the centre of the planet to support a column of mass against its weight.]

This turns out to be of order the mass of Jupiter; thus Jupiter is not only the largest planet in the Solar System, but one of the largest planets anywhere.

43.3   Nuclei without neutrons

Before the discovery of quantum mechanics in the mid 1920s the atomic nucleus was thought to consist of protons and electrons. (The neutron was not discovered until 1932.) Accordingly, a nucleus of mass number A and atomic number Z was thought to be made up of A protons to give it the correct mass and $(A - Z)$ electrons to give it the correct charge.

(a) Apply the uncertainty principle to an electron confined in the nucleus of a gold atom to estimate the minimum kinetic energy of the electron. The radius of a nucleus of mass number A is given by $r = 1.2 \times 10^{-15} A^{1/3}$ m.

(b) Evaluate the electrostatic potential energy of the electron at the surface of the gold nucleus. Hence explain why the discovery of quantum mechanics led to the abandonment of this theory.

(c) Why does this problem not arise for a nucleus composed of protons and neutrons?

Tutorial 44   Cross sections

When a pulse of N particles, of energy E, traverses a slab of material of thickness Δx, having n atoms per unit volume, the fraction $\Delta N/N$ of the incident particles that interact with atoms of the material is given by

$$\Delta N/N = -\sigma n \Delta x \qquad (4)$$

where σ is the total interaction cross section at energy E, $1/n\sigma$ is the mean free path of an incident particle in the material and $\Delta x \ll 1/n\sigma$ for the expression to be valid.

44.1   Supernova neutrinos

In February 1987 eleven neutrino events were registered in a time interval of about 12 s by the Japanese water detector Kamiokande. A similar burst of neutrinos was detected at the same time by the IMB water detector in the United States. Some hours later, the initial outburst of a supernova was seen in the Large Magellanic Cloud at a distance of 51 Kpc. Theory says that a type 2 supernova is caused by the collapse of the core of a massive star and that at least 99% of the gravitational binding energy released in the collapse is radiated away by the core in a pulse containing equal numbers of neutrinos and anti-neutrinos of all three flavours.

(a) Estimate the binding energy released in such a core collapse. At the onset of collapse the core of the star would have had a mass of about 1.4 solar masses and a radius of about 6000 km; the radius after collapse is about 10 km.

(b) Show that, from equation (54), the number of interactions may be written in the form

$$\Delta N = \sigma \times \text{time integrated flux of particle} \times \text{number of target particles.}$$

(c) Hence show that the number of neutrinos detected by Kamiokande is consistent with them coming from the supernova.

Water detectors register electron neutrinos via the reaction $\bar{\nu}_e + p \rightarrow e^+ + n$ on the free protons. This has a cross section of 9.23×10^{-42} $(E_\nu/10 \text{ MeV})^2$ cm^2, where E_ν, the neutrino energy, has an average value of about 14 MeV. The Kamiokande detector was operating with a fiducial mass of 2140 tonnes of ultra-pure water.

44.2   Yukawa's meson

Yukawa in 1934 proposed a field theory for the strong force which binds protons and neutrons together in the atomic nucleus. The mass of the particle mediating this field of force was dictated by the short range of the inter-nucleon force and was predicted to be somewhat over 200 electron masses. Then in 1937 the muon, with a mass of 207 electron masses, was discovered in cosmic rays and was immediately identified as Yukawa's mediator of the nuclear force. This question examines the evidence, from experiments conducted after the Second World War, which proved that the muon could not be Yukawa's particle.

A negatively charged muon can go into a Bohr orbit about the nucleus of an atom to form a muonic atom.

(a) Use the Bohr theory of the atom to show that the radius of the ground state orbit of the muon is 1/207 th of the radius of the smallest electron orbit.

(b) Why are we justified in applying the Bohr theory, which strictly speaking only works for hydrogen-like systems, to muon orbits in multi-electron atoms?

(c) For a high Z nucleus, for example lead, show that the radius of the ground state orbit is effectively inside the nucleus. The radius of a nucleus with mass number A is $1.2 \times 10^{-15} A^{1/3}$ m.

(d) Under these circumstances the muon undergoes the following interaction with a proton

$$\mu^- + p \rightarrow n + \text{something}$$

in a time of about 10^{-8} s. (The muon's lifetime is 2.2×10^{-6} s.) Estimate the interaction cross section of this process and hence explain why the muon cannot be Yukawa's particle.

(e) Identify the 'something' in the equation for the reaction above.

44.3   Too far flung for civilisation?

The density of stars in the central region of the Galaxy is of the order of 10^6 pc^{-3} and their speeds are about 200 km s^{-1}. Could an advanced civilisation develop in this part of the Galaxy?

Tutorial 45   Nuclear explosions

A nuclear explosion results when a fission chain reaction mediated by neutrons is initiated in a sufficiently large lump, called a *super-critical mass*, of nearly pure ^{235}U or ^{239}Pu. As a super-critical mass is inherently unstable it has to be created just prior to the explosion. This can be done either by the *gun method*, in which two sub-critical mass pieces are brought together by firing one piece into the other, or by the *implosion method*, in which the density of a single sub-critical mass lump is raised. The increase in density is achieved by detonating conventional explosives to compress a sphere of the fissionable material. The implosion method works because the critical mass decreases with increasing density, so sufficient compression will convert a sub-critical into a super-critical mass.

45.1 Critical mass

(a) Explain the existence of a critical mass for fissionable material.

(b) Explain why the mean free path of a neutron is inversely proportional to the density of the material. Hence show that the critical mass of fissionable material is inversely proportional to its density squared.

Note that the implosion method reduces the amount of fissionable material needed for criticality to be achieved and so reduces the size of the energy release. The critical mass can be further reduced by surrounding the fissionable material with a tamper. This is a substance which reduces the number of escaping neutrons by scattering some of them back.

(c) To initiate a chain reaction in a spherical mass of ^{235}U, the sphere must be large enough so that at least one of the 2.5 neutrons produced on average per fission is prevented from escaping across the surface. Argue that roughly the same proportion will escape from off-centre points as escape from the centre and therefore that for a quantitative estimate we need only consider neutrons escaping from the centre. Hence estimate the critical mass of a sphere of ^{235}U at normal density neglecting elastic scatterings. The density of uranium is 19 000 kg m^{-3}, the cross-section for fission by fast neutrons is 1.22×10^{-28}m^2 and the average number of neutrons per fission is 2.5.

45.2 Nuclear fireballs

(a) The prompt energy released in the fission of a nucleus of ^{239}Pu is about 180 MeV. Assuming that 50% of the nuclei in a sphere of compressed ^{239}Pu fission before criticality is lost through expansion, calculate the energy released per kilogram of fissionable material.

(b) This extremely rapid liberation of energy turns the plutonium sphere into an intensely hot plasma. Assuming that this plasma has become thermalised by the time it has expanded to the normal density of plutonium ($19\,000$ kg m^{-3}) estimate the temperature of the plasma and show that the major part of the energy is in the form of radiation. Take the mass of the plutonium sphere to be 2 kg.

(c) Hence, what is the pressure in the fireball at this instant?

(d) What is the typical energy of a photon coming from the fireball?

(e) A nuclear explosion is accompanied by a large electromagnetic pulse. Suggest a mechanism by which it might be generated.

Tutorial 46 Degenerate electron gases

Electrons have half integral spin and so obey the Pauli exclusion principle. This limits their distribution to a maximum of one per state. Even at the absolute zero of temperature therefore, the free electrons in a metal occupy a range of energy states. The number of states available to the electrons is h^{-3} per unit volume of configuration space per unit volume of momentum space per spin state, so the kinetic energy U of N electrons confined to a volume V is

$$U = \left(\frac{4\pi h^2 V}{5m}\right)\left(\frac{3n}{8\pi}\right)^{5/3}, \tag{5}$$

where n is the number density of the electrons. When the exclusion principle rather than the temperature determines the energies of the electrons, as is the case in metals below their melting points, the electrons are said to be degenerate.

46.1 A matter of compressibility

(a) Show that a degenerate electron gas exerts a pressure that depends on the number density of electrons.

In a metal this pressure force is balanced by the electrostatic interaction of the electrons with the ion cores of the lattice: the mutual potential energy between the electrons and the ion cores increases as their separation increases.

(b) What is the pressure P exerted by the degenerate electron gas in a metal given that the number density of electrons in a typical metal is 1.1×10^{29} m^{-3} ? Hence estimate the bulk modulus, $-V dP/dV$, of the metal.

(c) In tutorial 45 it is stated that the implosion method for initiating a nuclear explosion requires that a 2 kg sphere of Pu metal be compressed to about twice normal density by means of conventional explosives. Show that this degree of compression is feasible using high explosives. The energy yield of dynamite is about 4 MJ kg^{-1}.

46.2 Brown Dwarfs

For a cold body in the Solar System the more massive it is the larger its radius. This behaviour is characteristic of a body made up of un-ionised atoms. However, the pressure near the centre of a cold object with a mass somewhat greater than that of Jupiter is large enough to ionise atoms there (see tutorial 43). Such a body is supported against gravity by the degeneracy pressure of the free electrons in the interior. Bodies with masses intermediate between the mass of Jupiter and the mass of the smallest stars are called brown dwarfs. These bodies are insufficiently massive to initiate hydrogen burning (although at the upper end of the range they can burn ^{2}H).

Show that the radius of a brown dwarf is inversely proportional to its mass to the one third power (the inverse of the behaviour one might naively have expected). [Hint: minimise the total energy (gravitational potential energy + free electron kinetic energy) of the body. Make the approximation that its density is constant and assume it is composed of hydrogen.]

46.3   White Dwarfs

White dwarfs are stars in which nuclear burning has ceased, so they are also supported by degenerate electron pressure. Their masses range from about 0.08 solar masses, the minimum mass of a true star, to a maximum value known as the Chandrasekhar mass. In this question we investigate the reason for the existence of this upper limit to the mass of a cold body supported by the pressure of degenerate electrons. Equation (5) above assumes the degenerate electrons have non-relativistic energies. This assumption is valid for the degree of compression produced by gravity in brown dwarfs and in lower mass white dwarfs, but ceases to be valid for white dwarfs near to the Chandrasekhar limit. When the electron energies are highly relativistic equation (1) must be replaced by

$$U = 2\pi V ch (3n/8\pi)^{4/3}. \tag{6}$$

Show that with this expression there is an upper limit to the mass of a body which can be supported by degenerate electron pressure and obtain a value for this mass. Take the number of nucleons in the star to be twice the number of electrons. [Hint: Find the dependence of the total energy of the white dwarf on its volume.]

Part 2: Answers to Tutorials

Tutorial 1 Energy supply and storage

1.1 The Great Pyramid

(a) The volume of a pyramid is $\frac{1}{3} \times$ (area of base) $\times$ height, so the mass is

$$M = \frac{1}{3} \times (2700 \text{ kg m}^{-3}) \times (230.4 \text{ m})^2 \times (146.7 \text{ m}) = 7.0 \times 10^9 \text{ kg}.$$

The potential energy of a pyramid of height h and mass M is $\frac{1}{4} \times Mgh = \frac{1}{4} \times (7.0 \times 10^9 \text{ kg}) \times (9.8 \text{ m s}^{-2}) \times (146.7 \text{ m}) = 2.5 \times 10^{12}$ J.

(The factors of 1/3 in the volume and 1/4 in the potential energy can be derived by integration or omitted for an order of magnitude estimate.)

So the number of man days required to lift the stones is

$$\frac{2.5 \times 10^{12} \text{ J}}{2.4 \times 10^5 \text{ J man}^{-1}\text{day}^{-1}} \approx \boxed{10^7 \text{ man days.}}$$

This is obviously an underestimate of the labour involved because stones have to be cut and hauled to the site, and ramps have to be built to manoeuvre the stones into place. So we must multiply by at least a factor 2.

(b) In 20 years there are about 7300 days.

So over 20 years at least 1400 men would be required, or $\boxed{2800 \text{ men}}$ if we include the factor 2. Equivalently 11 200 men could have completed the job in 5 years. Thus a labour force of 100 000 men seems not to have been required (unless most of them were administrators).

1.2 Energy from burning

(a) The reaction is $C + O_2 \rightarrow CO_2 + 4\text{eV}$

1kg of ^{12}C contains

$$\frac{1000 \text{ g}}{12 \text{ g mole}^{-1}} \times (6.0 \times 10^{23} \text{ atoms mole}^{-1}) = 5 \times 10^{25} \text{ atoms.}$$

(i) So the energy from 1 kg of carbon (coal) excluding the mass of oxygen needed to burn it is

$(5 \times 10^{25} \text{ atoms}) \times (4 \text{ eV atom}^{-1}) \times (1.6 \times 10^{-19} \text{ ev J}^{-1}) = \boxed{3.2 \times 10^7 \text{ J kg}^{-1}}$.

(ii) With the oxygen a mole of CO_2 is $12 + 32 = 44$ g and the fraction of C is

12/44. As 1 kg of ^{12}C gives 3.2×10^7 J, 1 kg of C + O_2 gives

$$\frac{12 \text{ g}}{44 \text{ g}} \times (3.2 \times 10^7 \text{ J}) = \boxed{8.7 \times 10^6 \text{ J kg}^{-1}}$$

(b) High explosives carry their own oxygen so their specific energy yield must be of order $\boxed{10 \text{ MJ kg}^{-1}}$ i.e. the yield of high explosive is less than that from pure carbon, but, of course, the energy is released rapidly.

1.3   Battery power

(a) One cell supplies 1 W for 27 hours or

$(1 \text{ W}) \times (27 \text{ hr}) \times (3600 \text{ s hr}^{-1}) = \boxed{9.7 \times 10^4 \text{ J}}$.

This is $= \frac{9.7 \times 10^4 \text{ J}}{0.14 \text{ kg}} = 6.9 \times 10^5 \text{ J kg}^{-1}$.

(b) 1 unit = 1 kW hr = $10^3 \times 3600$ J = 3.6×10^6 J.

So the number of batteries needed to supply 1 kW hr = $(3.6 \times 10^6 \text{ J})/(9.7 \times 10^4 \text{ J}$ per battery$) \approx 37$. At £1.50 per battery this is $\boxed{£55.60}$, rather more than £0.08!

1.4   Alternative power for cars

(a) 1 kg of battery stores $(12 \text{ volts}) \times (60 \text{ A hr}) \times (3600 \text{ s hr}^{-1})/(12 \text{ kg})$ $= \boxed{0.2 \times 10^6 \text{ J kg}^{-1}}$, less than the non-rechargeable battery above and significantly less than the energy from burning carbon.

(b) 40 litres of petrol yields the same energy as $\frac{3}{2} \times 40$ kg of coal. So, from (1.2(a)(i)) the energy yield of 40 l of petrol is $(\frac{3}{2} \times 40 \text{ kg}) \times (3.2 \times 10^7 \text{ J kg}^{-1}) = 1.9 \times 10^9$ J.

Thus the equivalent number of batteries is $(1.9 \times 10^9 \text{ J})/[(12 \text{ kg}) \times (0.21 \times 10^6 \text{ J kg}^{-1})] = 754$, and their combined weight at 12 kg each is $\boxed{9048 \text{ kg}}$. This is 5 to 10 times the weight of a typical car. It is clear that conventional batteries cannot replace the internal combustion engine.

(c) The most favourable case for solar power would be a car covered in solar cells and operated at the equator. Then the power collected is the solar constant $\times$ the area of the car. Taking the area of a typical car to be 4.5×1.7 m^2 gives a power

$$(1360 \text{ W m}^{-2}) \times (4.5 \times 1.7 \text{ m}^2) \quad = \quad 1.04 \times 10^4 \text{ W}$$

$$\simeq \quad \boxed{14 \text{ hp}}. \quad (1 \text{ horsepower} = 746 \text{ W.})$$

This assumes 100% efficiency in converting light to electrical power. To achieve a 50% efficiency would be very good. Thus we might expect a maximum power output of a meagre 7 hp (compared with 60 - 120 hp for a modern family car.) Thus solar power is also not a realistic alternative to petroleum fuels.

1.5 The Pu battery of Cassini

(a) The energy released by the alpha decay of 1 kg of ^{238}Pu is

$$E = \left(\frac{6.0 \times 10^{26} \text{ atoms}}{238} \right) \times (5.5 \times 10^6 \text{ eV}) \times (1.6 \times 10^{-19} \text{ J eV}^{-1}) = \boxed{2.2 \times 10^{12} \text{ J}}$$

(b) The activity $A = \lambda N = \lambda N_0 e^{-\lambda t}$ decays s^{-1}.

The number of atoms at take-off

$$N_0 = (33 \text{ kg}) \times \left(\tfrac{6 \times 10^{26}}{238} \text{ atoms kg}^{-1} \right) = 8.3 \times 10^{25} \text{ atoms.}$$

Also

$$\lambda = \frac{\log_e 2}{T_{1/2}} = \frac{\log_e 2}{(88 \text{ yr}) \times (3 \times 10^7 \text{ s yr}^{-1})} = 2.6 \times 10^{-10} \text{ s}^{-1}.$$

So $A_0 = \lambda N_0 = 2.2 \times 10^{16}$ decays s^{-1}.

The power at take-off was

$$(2.2 \times 10^{16} \text{ decays s}^{-1}) \times (5.5 \times 10^6 \text{ eV decay}^{-1}) \times (1.6 \times 10^{-19} \text{ J eV}^{-1})$$
$$= 1.9 \times 10^4 \text{ W.}$$

This is reduced after 7 years by a factor $\exp\left(-\tfrac{7}{88} \log_e 2 \right)$ which is about 0.95. So the power on arrival is $\boxed{1.8 \times 10^4 \text{ W.}}$

Tutorial 2 Solar energy and power

2.1 Energy of a burning Sun

(a) The total energy output of the Sun over its lifetime is

$$\begin{aligned}
E &= \text{solar luminosity} \times \text{lifetime} \\
&= 4\pi (1.5 \times 10^{11} \text{ m})^2 \times (1360 \text{ W m}^{-2}) \times (4.5 \times 10^9 \text{ yr}) \times (3.2 \times 10^7 \text{ s yr}^{-1}) \\
&= \boxed{5.5 \times 10^{43} \text{ J.}}
\end{aligned}$$

(b) The mass of the Sun is 2×10^{30} kg.

The energy obtained by burning 2×10^{30} kg of coal and oxygen (from tutorial 1) is $(2 \times 10^{30} \text{ kg}) \times (8.7 \times 10^6 \text{ J kg}^{-1}) = 1.74 \times 10^{37}$ J

The lifetime of the coal Sun would be $\frac{1.75 \times 10^{37} \text{ J}}{5.5 \times 10^{43} \text{ J}} \times 4.5 \times 10^9$ yr $= 1.4 \times 10^3$ yr, a period shorter than recorded history.

Chemical processes liberate a few eV per atom so they all yield the same timescale.

2.2   Energy of a contracting Sun

(a) The energy released by contraction of a uniform star from an infinitely dispersed state down to a sphere of radius R is $\frac{3}{5}\frac{GM^2}{R}$. (The factor 3/5 comes from an integration over spherical shells of uniform density but can be omitted for an order of magnitude estimate. The numerical factor for a real star depends on the actual density distribution.) For the Sun this is

$$E_G = \frac{3 \times (6.7 \times 10^{-11} \text{ N kg}^{-2}\text{m}^2) \times (2 \times 10^{30} \text{ kg})^2}{5 \times (7 \times 10^8 \text{ m})} = \boxed{2.3 \times 10^{41} \text{ J.}}$$

(b) The lifetime at current luminosity is

$$t = \frac{2.3 \times 10^{41} \text{ J}}{(3.9 \times 10^{26} \text{ W}) \times (3.2 \times 10^7 \text{ s yr}^{-1})} = \boxed{1.8 \times 10^7 \text{ yr.}}$$

In fact, the contracting star would be centrally condensed (the density would be higher towards the centre) which would increase the lifetime, but not by enough to exceed the geological age of the Earth, which was then estimated to be as high as several hundred million years. (Although it still comfortably exceeds Bishop Usher's biblical chronometer which put the creation of the Earth at 4004 BC.)

2.3   Helium creation

The total energy output of the Sun over its lifetime from 2.1(a) is 5.5×10^{43} J. So the number of ^{4}He nuclei formed is

$$\frac{5.5 \times 10^{43} \text{ J}}{(26.7 \times 10^6 \text{ eV}) \times (1.6 \times 10^{-19} \text{ J eV}^{-1})} = 1.3 \times 10^{55} \text{ nuclei.}$$

The mass of each ^{4}He nucleus is about $4m_p$ so the total mass of ^{4}He formed is

$$(1.3 \times 10^{55} \text{ nuclei}) \times (4 \times 1.67 \times 10^{-27} \text{ kg nucleus}^{-1}) = \boxed{8.9 \times 10^{28} \text{ kg.}}$$

As a percentage of the Sun's mass this is $\frac{8.9 \times 10^{28} \text{ kg}}{2 \times 10^{30} \text{ kg}} \times 100 = 4.5\%$.

2.4   Neutrinos from the Sun

(a) (i) The luminosity of the Sun is 3.9×10^{26} W. The number of ^{4}He nuclei produced per second is

$$\frac{\text{luminosity}}{\text{energy release per } {}^4\text{He}} = \frac{3.9 \times 10^{26} \text{ W}}{(26.7 \times 10^6 \text{ eV}) \times (1.6 \times 10^{-19} \text{ J eV}^{-1})}$$
$$= 9.1 \times 10^{37} \text{ s}^{-1}$$

The number of neutrinos emitted per second is twice this i.e. 1.8×10^{38} s^{-1}.

The flux at the Earth is $\frac{1.8\times10^{38}\text{ s}^{-1}}{4\pi\times(1.5\times10^{11}\text{ m})^2} = \boxed{6.4\times10^{14}\text{ m}^{-2}\text{s}^{-1}}$.

(ii) The area of the human body is, say, about 0.5 m^2, so the total number flux through each of us is $\sim 3\times10^{14}$ s^{-1}.

The energy flux is therefore $(3\times10^{14}\text{ s}^{-1})\times(0.3\times10^6\text{ eV})\times(1.6\times10^{-19}\text{ J eV}^{-1}) \approx \boxed{14\text{ J s}^{-1}}$.

The resting human body radiates about 75W so this is an appreciable amount of energy. Yet the neutrinos interact so weakly that is it unlikely that a single solar neutrino will be absorbed by the body in a lifetime.

(b) The neutrino energy flux is not included in the solar constant as it does not register on solar bolometers. The energy flux of $(6.4\times10^{14}\text{ m}^{-2}\text{s}^{-1})\times(0.3\times10^6$ eV$)\times(1.6\times10^{-19}\text{ J eV}^{-1}) = 30.7$ W m^{-2} in neutrinos represents a fraction $\frac{30.7}{1360+30.7}$ or 2.2 % of the total. Since we used the solar constant to get the total solar energy over its lifetime in question (2.1) our estimate of the helium production in question (2.3) is low by $\boxed{2.2\%.}$

Tutorial 3   Newtonian games

3.1   Athletic records

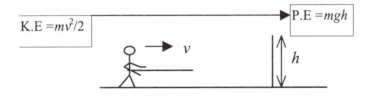

Figure 16. The vaulter converts kinetic to potential energy

The exertions of the vaulter gain him mgh in potential energy in return for the $mv^2/2$ in kinetic energy with which he approaches the vault. Take the height vaulted to be $h = 7$ m (not 7.5 m, since the centre of gravity of the vaulter as he approaches the bar is above the ground and he arches over the bar using his arms). Thus, he is approaching the bar at speed v given by

$$mgh = mv^2/2.$$

i.e. $v^2 = 2\times(9.8\text{ m s}^{-2})\times(7\text{ m})$ or $v\simeq12$ ms^{-1}. If this is the mean speed of a sprinter over the 100 m then the time is $100/12 = \boxed{8.3\text{ s.}}$

The main source of error is presumeably in assuming a 100% efficient conversion of energy from kinetic to potential for the vaulter. Errors are worked out efficiently by logarithmic differentiation. Since $mv^2/2 = E$ then $2\log_e v + \log_e m/2 = \log_e E$. We assume that v and E are subject to errors δv and δE but that m is known accurately so $\delta m = 0$. Then differentiation gives $2\delta v/v = \delta E/E$. The error in E is likely to be more than $\delta E/E = 1\%$ and less than 100% so call it 10%. This would mean v has been underestimed by 5%, and hence the time might be less by 5%.

3.2   Hazardous high jumps

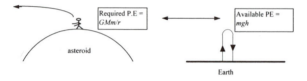

Figure 17. Jumping off an asteroid

The world high jump record is 2.4 m, so the earthling could raise his centre of gravity by no more than 2 m in the Earth's gravity. The energy available to launch the olympian into space is therefore mgh, where m is the mass of a human and $h = 2$ m, the height he can raise himself on Earth. The gravitational energy needed to escape to infinity against gravity from a spherical body of mass M and radius r is GMm/r. For an asteroid, assumed spherical, of mass M, the minimum radius r is given by

$$GMm/r = mgh.$$

Asteroids can be assumed to have a density of $\rho = 2 \times 10^3$ kg m^{-3}. Thus, $G(\frac{4}{3}\pi r^3 \rho)/r = gh$ or

$$r = \left(\frac{3}{4}\frac{gh}{\pi G\rho}\right)^{1/2} = \left(\frac{3 \times (9.8 \text{ m s}^{-2}) \times (2 \text{ m})}{4\pi \times (6.7 \times 10^{-11} \text{ N m}^2 \text{ kg}^{-2}) \times (2 \times 10^3 \text{ kg m}^{-3})}\right)^{1/2}$$

$$= \boxed{5.9 \text{ km}}.$$

Thus, it is safe to hold the games on the moons of Mars, Phobos and Deimos, which have radii 12 km and 24 km.

3.3 Improve your golf

(a) The club transmits the force of impact to the golfer. But the force is transmitted at the speed of sound, about 1000 m s^{-1} in a solid, so it takes only about 1 ms for the impulse to travel up the golf club. The travel time up the shaft is longer than the contact time. Thus, by the time the impulse has travelled up the shaft the ball is on its way and there is no possibility of transmitting any influence between the golfer and the ball. Therefore, we can treat the problem as a collision between the club head and the ball and ignore the golfer. If you hit the ground instead of the ball the difference is obvious. The club transmits the forces between the Earth and the golfer in a time which is short compared with the impact time. This impact therefore takes place, in effect, between the Earth and the golfer.

This point has been demonstrated with a club which has a hinge in its shaft just above the clubhead. The shots hit with this club go almost the same distance as shots hit with a normal club. (See *The Search for the Perfect Swing,* A. Cochran & J. Stobbs)

There is another point relating to the fact that the golfer accelerates the club throughout the impact. We do not therefore appear to have an isolated system and we should not conserve momentum. However, if the impact takes $< 10^{-3}$ s and the acceleration is $10g$, the change in speed of the club is 10 cm s^{-1} which is negligible. In practice therefore we can consider the system of ball and clubhead as isolated (i.e. subject to no external forces).

(b) There are several equivalent ways of looking at this. First consider the impact of an elastic ball with a stationary object, such as the ground. The ball is initially compressed, then springs back into shape and rebounds. The time of contact is therefore equal to half a period of oscillation of the ball, which is independent of the amplitude of the oscillation (at least for small deformations). Strictly speaking, there is also a small deformation of the ground which communicates a small momentum to the Earth. We get a similar picture of the impact of the golf club and ball when viewed in the frame of the club head, so here too the contact time is independent of the weight of the shot.

Alternatively, we can say that the ball cannot move off as an entity until the impact has been communicated from the struck face to the front face of the ball and back to the club head. This occurs by means of compression (or sound) waves in the ball. Thus the time of contact is of order the time for a sound wave to traverse the ball, which is independent of the amplitude of the wave, and hence independent of the strength of the impact. To order of magnitude the sound crossing time and the period of oscillation of an elastic object are the same, so these are two ways of looking at the same thing.

(c) Let the club head have mass M and the ball mass m. ($M = 200$ g and $m = 46$ g, so $m/M \sim 1/4$.) Let the club have speed u before impact and $\bar{u}$ after, and the ball velocity v. Conserving momentum gives

$$Mu = M\bar{u} + mv;$$

conserving energy gives

$$Mu^2 = M\bar{u}^2 + mv^2.$$

Thus, eliminating $\bar{u}$ gives

$$v = 2u/(1 + m/M).$$

Increasing M to infinity (from $4m$) increases v by only about 20%, so increasing the clubhead mass is hopelessly inefficient. On the other hand increasing u increases v in direct proportion.

(d) Increasing the shaft length, ceteris paribus ('other things being equal'), also increases the speed in proportion.

A note to non-golfers: increasing one's natural swing speed reliably is difficult in practice. The third strategem is quite effective however and is sometimes adopted by professional golfers.

3.4   Snooker insights

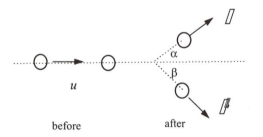

u

before   after

Figure 18. Impact of snooker balls

(a) For a ball of mass m, conserving momentum parallel to the motion of the incoming ball in figure (18) gives

$$mu = mv\cos\alpha + mv'\cos\beta, \qquad\qquad (7)$$

and for momentum perpendicular to this

$$0 = mv\sin\alpha - mv'\sin\beta. \qquad\qquad (8)$$

Finally, conserving energy gives

$$\frac{1}{2}mu^2 = \frac{1}{2}mv^2 + \frac{1}{2}mv'^2. \tag{9}$$

Then $(7)^2 + (8)^2$ gives

$$\begin{aligned} u^2 &= (v\cos\alpha + v'\cos\beta)^2 + (v\sin\alpha - v'\sin\beta)^2 \\ &= v^2 + v'^2 + 2vv'\cos\alpha\cos\beta - 2vv'\sin\alpha\sin\beta \end{aligned}$$

which, using (9), gives

$$0 = \cos\alpha\cos\beta - \sin\alpha\sin\beta = \cos(\alpha + \beta).$$

So $\alpha + \beta = \pi/2$, as required.

(b) In practice the main complicating factor is the rotational energy of the balls.

Tutorial 4   Damage limitation

4.1   Elastic and inelastic collisions

(a) Energy goes into deforming the club and the ball. Once contact is over the club and ball are free to oscillate about their equilibrium shapes. This oscillation is damped by internal friction and dissipated as heat.

(b) The rubber bullet will rebound, the aluminium one will embed itself in the wooden block, so the momentum transfer is twice as much in the former case. Thus the rubber bullet is the one most likely to knock over the target. On the other hand the aluminium bullet dissipates most of its energy in the block so will cause the greater damage.

(c) Treat the tackle as an inelastic collision and assume that in both cases the player tackled has equal and opposite momentum. In both cases the player tackled will be brought to rest, so the two tackles are equally effective. On the other hand the smaller player (the scrum half) has more energy which has to be dissipated in the impact. So his tackle should be the more painful.

Note how the conclusion is made obvious by choosing an apparently special case. By altering one's frame of reference (which cannot alter the physics) the player to be tackled can be given any momentum, so the conclusion stands also in general. In fact, from what we remember of school rugby, it was always a good idea to keep out of the way of the bigger guys. This means our assumption that the scrum half and the forward develop equal momentum is probably unrealistic.

4.2   Brute strength or physics?

As the hammer does not break the slab take the collision of hammer and slab to be elastic. The explanation hinges on the amount of energy transferred between the hammer and the slab in the collision. If the hammer has mass m and the slab mass M, we shall show below that the slab acquires a kinetic energy of order $(m/M)\times$ the energy of the hammer. The capacity to cause damage has therefore been reduced by this ratio.

To crack ribs requires a minimum energy which the slab does not acquire. Instead it compresses the rib cage to a point below the elastic limit of the bone and recoils without causing damage. The trick is clearly to use as massive a slab as possible and not too large a hammer! It also helps to have a strong rib cage. The mathematics should tell you precisely how to arrange things safely - but we have not verified it experimentally, so please do not try this version. There is an equivalent party trick with a slab on top of a glass which is safer.

To prove the result refer to question 3.3. By analogy, if u is the speed of the hammer before impact, the speed of the slab after impact is $v = 2u/(1 + M/m)$. For $M \gg m$ this is $v \sim 2mu/M$. The kinetic energy of the slab is $(M/2)(2mu/M)^2 = (4m/M)(Mu^2/2)$ as was to be proved. So for $M > 4m$ there is a reduction in energy. A second effect of the slab is that, unlike the hammer, it transfers the energy to the whole rib cage rather than just one or two ribs.

Tutorial 5   The power of force

5.1   Motion under constant power

(a) Let the force acting be F. Since power is the rate of doing work we have $P = Fv$. Thus $\boxed{F = P/v}$.

(b) This question is about not jumping to conclusions! The rate of change of momentum is the force F.

Thus if P is constant, the rate of change of momentum, $F = P/v$ is changing. (And since $F = ma$, the acceleration is not constant.).

(c) On the other hand the rate of change of energy is the rate at which work is being done, which is constant for a constant power.

(d) A constant rate of change of energy implies the energy increases linearly with time i.e. $v^2 \propto t$.

A constant acceleration (i.e rate of change of velocity) implies v increases linearly with t, i.e. $v \propto t$.

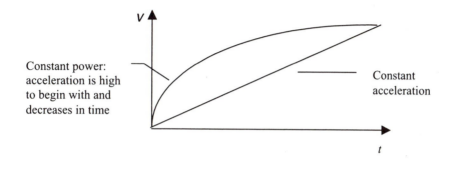

Figure 19. Velocity against time

5.2 The Channel tunnel shuttle

(a) The acceleration $\boxed{a = P/mv}$.

(b) To find the velocity as a function of time we integrate the equation of motion for the given force law. We have

$$dv/dt = P/mv \text{ or } vdv/dt = P/m \text{ and therefore } mv^2/2 = Pt,$$

since $v = 0$ at $t = 0$. Hence the time T at which running speed is reached is

$$T = mv^2/2P = \frac{(2500 \times 10^3 \text{ kg})}{2 \times (11 \times 10^6 \text{ W})} \times \left(\frac{140 \times 10^3 \text{ m hr}^{-1}}{3600 \text{ s hr}^{-1}}\right)^2 = \boxed{2.9 \text{ mins}}$$

(c) To find distance as a function of time we integrate the velocity,

$$\frac{dx}{dt} = v = \left(\frac{2P}{m}\right)^{1/2} t^{1/2}, \tag{10}$$

which gives

$$x = \left(\frac{2}{3}\right) \left(\frac{2P}{m}\right)^{1/2} t^{3/2}$$

since $x = 0$ at $t = 0$. This yields

$$x = \frac{2}{3} \times \left(\frac{2 \times (11 \times 10^6 \text{ W})}{2500 \times 10^3 \text{ kg}}\right)^{1/2} \times (2.9 \times 60 \text{ s})^{3/2} = \boxed{4.5 \text{ km}}$$

(d) The average value of the force over time is given by

$$\overline{F} = \frac{\int_0^T F(t)dt}{T}$$

Here

$$F(t) = \frac{P}{v} = \frac{P}{(2P/m)^{1/2}t^{1/2}}.$$

The integral gives $\overline{F} = (2mP/T)^{1/2}$.

This constant force would provide a constant acceleration $a = (2P/mT)^{1/2}$.

To get the travel time and distance we can use the constant acceleration formula: $t = v/a$. Substituting for v at time T from part (c), as given in equation (10), we get

$$t = \frac{(2P/m)^{1/2} T^{1/2}}{(2P/mT)^{1/2}} = T = 2.9 \text{ mins.}$$

This equality of times is not surprising because we are using a time average of the force.

We also get

$$\begin{aligned} x &= v^2/2a = \frac{(2P/m)T}{2(2P/mT)^{1/2}} = \left(\frac{P}{2m}\right)^{1/2} T^{3/2} \\ &= \left(\frac{11 \times 10^6 \text{ W}}{2 \times 2500 \times 10^3 \text{ kg}}\right)^{1/2} \times (2.9 \times 60 \text{ s})^{3/2} \\ &= \boxed{3.4 \text{ km}} \end{aligned}$$

It takes further to reach the running speed because the acceleration in the case of constant power is initially much larger (compare the areas under the graphs in figure 19).

5.3   Drag racers

For a drag racer of power P and mass m, the velocity is given by

$$mv\frac{dv}{dt} = P.$$

Since the terminal velocity is attained at a fixed distance we want to consider the relation between v and distance travelled, x. We therefore rewrite the equation using $dv/dt = v\,dv/dx$ as

$$mv^2\frac{dv}{dx} = P.$$

Separating the variables and integrating gives

$$\frac{1}{3}mv^3 = Px$$

from which it follows that, for a given distance, $v^3 \propto P$ or $\boxed{v \propto P^{1/3}.}$

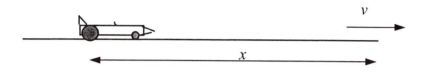

Figure 20. Drag racing

Tutorial 6   The centre of momentum frame

6.1   A traffic accident

In collision (i) not all the $\frac{1}{2}mv^2$ of kinetic energy is available to cause damage since conservation of momentum inplies both cars must be moving after the collision.

In collision (ii) the cars come to rest after the colllision so the kinetic energy available is $2 \times \frac{1}{2}m(v/2)^2 = \frac{1}{4}mv^2$.

The amount of damage cannot depend on the frame from which we view the event. So in (i) look at the collision in the centre of mass frame: it is exactly the same as collision (ii)! Thus the damage must be equal.

6.2   Forbidden processes

(a) Consider the putative absorption of the photon by the lone electron. The process is shown in the laboratory frame in figure (22) and transformed to the CM frame in figure (23).

In the CM frame it is clear that momentum is conserved but relativistic energy is not since, comparing the energy before and after, $h\nu' + \gamma mc^2 > mc^2$ as $\gamma > 1$.

(b) When an isolated atom absorbs a photon its rest mass changes. This is not possible for the electron since, being elementary, it has no excited states.

(c) In the laboratory frame the putative process is shown in the figure.

In the final state the total momentum is zero in the centre of mass frame. However, for the initial state, there is no frame in which the momentum is zero because a single photon cannot have a rest frame. It is therefore obvious that momentum cannot be conserved so the process cannot occur.

6.3   Creating the Z^0

(a) Let the lab. frame S and the CM frame S′ be inertial frames in standard configuration. Let frame S′ have speed v as seen from S. In the laboratory frame

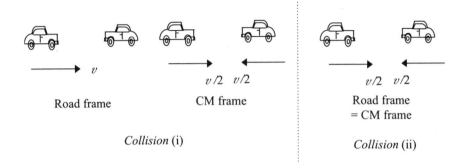

Road frame   CM frame   Road frame = CM frame

Collision (i)   *Collision* (ii)

Figure 21. Two collisions: (i) with one car stationary and (ii) with both in motion

S the moving proton has momentum p_1 and energy $E_1 = 500 \text{ GeV} + mc^2$. The target has momentum $p_2 = 0$, and energy $E_2 = mc^2$.

In the CM frame S' the protons have equal and opposite momenta, $p'_2 = -p'_1$ and equal energies $E'_1 = E'_2$. In the CM frame all the energy $E' = E'_1 + E'_2$ is available to create new particles. The question can be answered therefore if we calculate E' in terms of the given energy E. Clearly we have to relate quantities in the S' frame to those in the S frame, which we do by a Lorentz (energy-momentum) transformation. We have

$$E'_2 = \gamma(E_2 - vp_2) = \gamma E_2$$

so

$$E' = E'_1 + E'_2 = 2E'_2 = 2\gamma E_2 = 2\gamma mc^2. \tag{11}$$

Now we have to eliminate γ. Using the energy momentum transformation again, but now from S' to S,

$$E_1 = \gamma(E'_1 + vp'_1),$$
$$E_2 = \gamma(E'_2 + vp'_2) = \gamma(E'_2 - vp'_1),$$

and adding these two equations gives

$$E = E_1 + E_2 = \gamma(E'_1 + E'_2) = \gamma E'. \tag{12}$$

Eliminating γ between equations (11) and (12) gives finally

$$E' = (2mc^2 E)^{1/2}. \tag{13}$$

The energy available for the creation of particles is thus the geometric mean of the laboratory frame energies of the colliding particles. For a 500 GeV proton this gives

$$E' = (2 \times 0.938 \text{ GeV} \times 500 \text{ GeV})^{1/2} = \boxed{30.6 \text{ GeV}}.$$

(b) In equation (13) we require that E' be the energy of a Z^0. The required beam

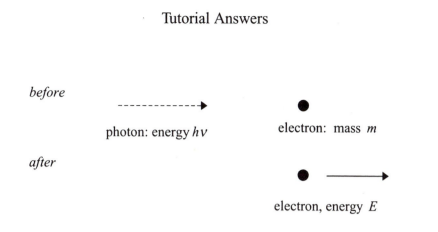

Figure 22. A photon and electron in the laboratory frame

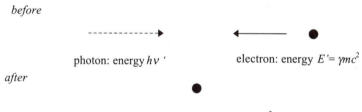

Figure 23. Photon and electron in the CM frame

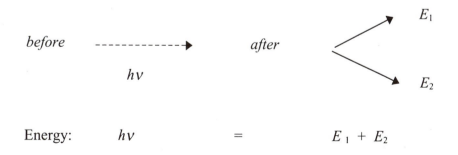

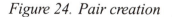

Figure 24. Pair creation

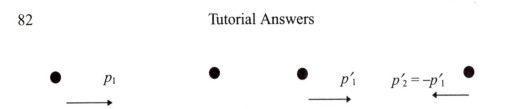

Figure 25. The collision of protons seen from the lab. frame on the left and the CM frame on the right.

energy E is therefore

$$E = \frac{E'^2}{2mc^2} = \frac{91^2}{2 \times 0.938} = \boxed{4414 \text{ GeV}}.$$

Tutorial 7   Round and round

7.1   A fairground rotor

(a) When the floor is removed the friction from the wall must support the riders against gravity. Assume that the riders are point masses (so the centre of mass of each is at radius $d/2$), and let the coefficient of friction be μ. Balancing forces vertically gives

$$\mu \frac{d}{2} \Omega^2 = g.$$

Taking $\mu = 1$ (its maximum value) gives $\boxed{\Omega = 2.2 \text{ rad s}^{-1}}$ or about 1 revolution every 3 s.

(b) A larger person has a smaller d, so Ω has to be slightly larger.

7.2   The wall of death

The figure shows the forces acting on the motorcyclist.
Equating forces horizontally and vertically gives

$$\frac{mv^2}{r} = N \tag{14}$$

and

$$\mu N = mg. \tag{15}$$

Eliminating N we get

$$\frac{gr}{v^2} = \mu.$$

(a) Since μ cannot exceed 1 the minimum speed is

$$v = \sqrt{gr} = \sqrt{9.8 \text{ m s}^{-2} \times 5 \text{ m}} \simeq \boxed{7 \text{ m s}^{-1}}.$$

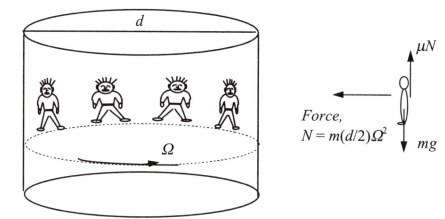

Figure 26. Forces in a fairground rotor

(b) Equating torques about A gives
$$\frac{mv^2}{r} l \sin\theta = mgl \cos\theta$$
where l is the distance of the centre of gravity from the wall. Thus
$$\boxed{\tan\theta = gr/v^2.}$$

(c) From part (a) the minimum speed is $v = \sqrt{gr}$ for which $\boxed{\theta = 45^0}$ and this is the largest angle.

(d) Notice that the mass does not appear in the above formulae so apparently the pillion passenger makes no difference. In fact, the pillion passenger will shift the centre of mass of the bike away from the wall and hence reduce the radius r of curvature of the path of the centre of mass. This will increase the minimum velocity by a small amount.

7.3 Rotating fluid

In an inertial frame, consider an element of fluid, mass m, at the surface as in the figure. Since the fluid surface is at the ambient pressure there can be no tangential pressure gradient, so the the force of gravity must provide the centripetal force,
$$mg \sin\theta = m(r\omega^2 \cos\theta).$$
Thus, $\tan\theta = r\omega^2/g$. But $\tan\theta = dz/dr$. Hence,
$$\frac{dz}{dr} = \frac{\omega^2 r}{g}.$$

Integrating, we get $r^2 = 2gz/\omega^2 +$ constant, the equation of a paraboloid in the r-z plane.

Tutorial 8   About turning

8.1   Walking a tightrope

(a) Let I_{total} be the combined moment of inertia of the walker and the pole and let G be the couple acting on them. Then at angle θ to the vertical the equation of motion of the system is

$$G = I_{total}\ddot{\theta}. \tag{16}$$

From figure 29, the torque is

$$G = Mg\frac{l}{2}\sin\theta - mflg\sin\theta$$

$$= (M + 2fm)\frac{lg}{2}\sin\theta.$$

The moment of inertia of the walker about the point of contact with the rope is

$$I_{walker} = \frac{1}{3}Ml^2.$$

The moment of inertia of the pole about the rope (ignoring any sag) by the parallel axes theorem is

$$I_{pole} = \frac{1}{12}mL^2 + m(fl)^2,$$

and $I_{total} = I_{walker} + I_{pole}$.

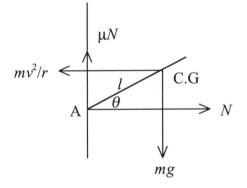

Figure 27. Force diagram for the wall of death

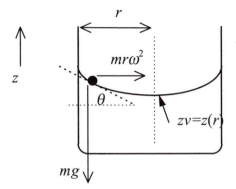

Figure 28. Rotating fluid viewed from an inertial frame

(b) For a small angle θ, we approximate $\sin \theta \simeq \theta$, so (16) becomes
$$\ddot{\theta} = \frac{(M + 2mf)\,lg}{I_{total}}\frac{\theta}{2}.$$
The solution
$$\theta = A \sinh \Lambda t$$
with
$$\Lambda = \left(\frac{(M + 2fm)lg}{2I_{total}}\right)^{1/2}$$
satisfies the condition $\theta = 0$ at $t = 0$. For small t we can approximate $\sinh \Lambda t \simeq \Lambda t$ so θ grows initially as $A\Lambda t$. The time constant (the logarithmic rate of growth, roughly a measure of the time taken to grow by A) is $\theta/(d\theta/dt) = 1/\Lambda$ for $t \gg \lambda$.

(c) The walker without the pole (equivalent to putting $m = 0$) has a time constant
$$\lambda^{-1} = \left(\frac{Mgl}{2I_{walker}}\right)^{-1/2}.$$
The ratio of time constants with and without the pole is therefore given by
$$\frac{\lambda^2}{\Lambda^2} = \frac{M}{(M + 2fm)}\left[1 + \frac{mL^2}{4Ml^2} + \frac{3m(fl)^2}{Ml^2}\right].$$
Evaluating this for a walker of mass 60 kg and height 1.7 m in the case where the centre of gravity of the pole is at the wire, $f = 0$, gives
$$\frac{\lambda^2}{\Lambda^2} = 1 + \frac{mL^2}{4Ml^2} = 1 + \frac{(14\text{ kg}) \times (12\text{ m})^2}{4 \times (60\text{ kg}) \times (1.7\text{ m})^2} = 3.9,$$
or $\boxed{\lambda/\Lambda \approx 2}$. Thus the time to respond is roughly doubled.

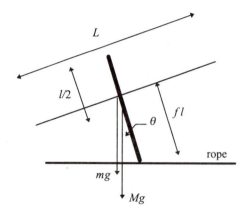

Figure 29. Mass model for the tightrope walker and pole

(d) For a massless pole with a mass $m/2$ at each end

$$I_{\text{pole}} = 2 \times \left(\frac{m}{2}\right) \left(\frac{L}{2}\right)^2 = \frac{mL^2}{4}.$$

So, with $f = 0$, we now get

$$\frac{\lambda^2}{\Lambda^2} = 1 + \frac{3mL^2}{4Ml^2} = 1 + 3 \times 2.9 = 9.7.$$

or $\boxed{\lambda/\Lambda \approx 3}$.

(e) The walker needs to be short and light, but strong to carry and handle the pole.

8.2   Tracking the angular momentum

(a) If the Sun is taken to have uniform density, mass M and radius R, and to be rotating uniformly with angular velocity ω, the angular momentum is

$$H_\odot = I\omega \simeq \frac{2}{5}MR^2 \times \frac{v}{R}$$

where v is the velocity of the surface at the equator. Hence, for the young Sun,

$$\begin{aligned}
H_\odot &= \frac{2}{5} \times (2 \times 10^{30} \text{ kg}) \times (7 \times 10^8 \text{ m}) \times (200000 \text{ ms}^{-1}) \\
&= \boxed{1.1 \times 10^{44} \text{ kg m}^2\text{s}^{-1}}.
\end{aligned}$$

The surface velocity of the present Sun is

$$\frac{2\pi R}{T} = \frac{2\pi}{(27 \text{ days}) \times (24 \times 3600 \text{ s day}^{-1})} \times (7 \times 10^8 \text{ m})$$

$$= \quad 1.9 \text{ km s}^{-1}.$$

The corresponding estimate of the present angular momentum is therefore about $1/100^{\text{th}}$ of that of the protosun.

(b) The present solar wind is about 2×10^{-14} $M_\odot$ yr^{-1}. Therefore the solar wind of the protosun is about 2×10^{-11} $M_\odot$ yr^{-1}. For simplicity assume that the solar wind is confined to the equatorial plane. This will over-estimate the angular momentum loss since the solar wind emitted at higher latitudes takes away less angular momentum. If the solar wind decouples from the Sun at a distance of 1 a.u. then the angular momentum lost by the early Sun over a period of the order of $t = 10^8$ years is

$$
\begin{aligned}
H \quad &= \quad \dot{M} v R t \\
&= \quad (2 \times 10^{-11} \ M_\odot \ \text{yr}^{-1}) \times (2 \times 10^{30} \ \text{kg} \ M_\odot^{-1}) \times (2 \times 10^5 \ \text{m s}^{-1}) \\
&\quad \times (1.5 \times 10^{11} \ \text{m}) \times 10^8 \ \text{yr} \\
&= \quad 1.2 \times 10^{44} \ \text{kg m}^2 \ \text{s}^{-1}.
\end{aligned}
$$

This is of the appropriate order of magnitude to carry away the excess angular momentum.

The corresponding mass loss is

$$(2 \times 10^{-11} \ M_\odot \ \text{yr}^{-1}) \times (2 \times 10^{30} \ \text{kg} \ M_\odot^{-1}) \times 10^8 \ \text{yr} = 4.0 \times 10^{27} \ \text{kg},$$

about 0.2 % of the present mass.

Tutorial 9   Rocket science

9.1   Steam powered space flight

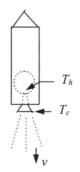

Figure 30. A schematic rocket motor

(a) A rocket motor is a heat engine which converts some of the thermal energy of its working substance into mechanical work. The most efficient heat engine is a Carnot engine which has an efficiency

$$\frac{\text{work out}}{\text{heat in}} = 1 - \frac{T_c}{T_h}$$

where T_h is the temperature of the hot source from which heat is extracted and T_c is the temperature of the cold sink into which waste heat is dumped. So a Carnot engine working between finite temperatures cannot be 100% efficient, and hence nor can any other heat engine.

Most of the waste heat will be in the form of thermal energy of the exhaust gases - they are still hot when they leave the exhaust nozzle.

(b) The thermal efficiency of an ideal rocket is

$$\eta = \frac{\text{kinetic energy of exhaust gases}}{\text{total chemical energy released by burning}}$$

Conservation of energy for a mass ΔM ejected gives

$$\Delta M c_p T_h = \Delta M c_p T_c + \frac{1}{2} \Delta M v_e^2,$$

where T_h and T_c are the temperatures of the hot gas and the exit gas respectively and c_p its specific heat at constant pressure.

Thus,

$$\eta = \frac{\frac{1}{2} \Delta M v_e^2}{c_p T_h} = \frac{c_p T_h - c_p T_c}{c_p T_h} = 1 - \frac{T_c}{T_h}.$$

Evidently the ideal rocket has the same efficiency as the cyclical Carnot engine. This is because (i) like the Carnot engine the ideal rocket works between two fixed temperatures: the combustion temperature of the burning gas and the final temperature of the exhaust gas, and (ii) the heat extracted at the higher temperature less the heat rejected at the lower all goes into useful work. The ideal Carnot engine achieves this in a cyclical manner and the ideal rocket in a non-cyclical one.

(c) To evaluate the efficiency from the data given we need to express η in terms of the initial and final pressures. To do this we need

the perfect gas law

$$PV = \Re T$$

and the adiabatic condition

$$PV^\gamma = \text{constant,}$$

from which

$$\frac{T_c}{T_h} = \left(\frac{P_c}{P_h}\right)^{\frac{(\gamma-1)}{\gamma}}$$

So we have

$$\eta = 1 - \left(\frac{P_c}{P_h}\right)^{\frac{(\gamma-1)}{\gamma}} = 1 - \left(\frac{1}{100}\right)^{\frac{0.25}{1.25}} = \boxed{0.60}.$$

(d) From the conservation of energy $\frac{1}{2}\Delta M v_e^2 = \eta\times$ the energy released by burning. So, per kilogram,

$$v_e^2 = 2\eta \times (13.3 \times 10^6 \text{ MJ kg}^{-1})$$

which, for $\eta = 0.6$ gives $\boxed{v_e = 3994 \text{ m s}^{-1}}$

9.2   Single stage to orbit

From the given integrated rocket equation and taking $\dot{m} = $ constant and exhaust speed v_0,

$$u = v_0 \log_e \frac{m_0}{m} - \frac{g}{\dot{m}}(m_0 - m),$$

where m_0 is the initial and m the final mass. To obtain $\dot{m}$ we are given the thrust,

$$v_0 \dot{m} = f m_0 g$$

with $1.5 < f < 2$. So, from the given data

$$u = 4500 \ln 10 - 0.9 \times 4500/f,$$

and hence $0.77 \text{ km s}^{-1} < u < 8.34 \text{ km s}^{-1}$.

The velocity required in low stable Earth orbit of radius R is $v_{orb} = \left(\frac{GM}{R}\right)^{1/2}$. Taking the altitude to be 300 km (for which the approximation $g = $ constant used above is adequate) gives

$$v_{orb} = \left(\frac{(6.67 \times 10^{-11}) \times (5.97 \times 10^{24})}{(6378 + 300) \times 10^3}\right)^{1/2} = 7.72 \text{ km s}^{-1}.$$

So the upper value of f is just sufficient to allow a rocket to achieve low Earth orbit. However, the escape velocity, $v_{esc} = (2GM/R)^{1/2} = 11.2 \text{ km s}^{-1}$, is definitely beyond the capabilities of such a single stage rocket.

Tutorial 10   Diluting gravity

10.1   Rolling down inclines

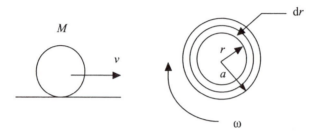

Rolling disc

(a) Linear momentum = $\boxed{Mv}$

(b) Assuming the material of the disc is uniform the mass of a column of given cross section is proportional to the cross sectional area. Thus, the mass of the annulus is

$$dM = \frac{2\pi r \, dr}{\pi a^2} M,$$

where M is the total mass. And the angular momentum of the annulus is

$$dMr^2\omega = \frac{2\pi r \, dr}{\pi a^2} Mr^2\omega$$

So the total angular momentum of the disc is

$$L = \frac{2M\omega}{a^2} \int_0^a r^3 dr = \boxed{\tfrac{1}{2}Mva},$$

since $v = a\omega$ is the condition for rolling.

(c) The total energy is made up of the kinetic energy of linear motion and that of rotation. The rotational energy can be calculated by summing up that for each annulus:

$$E_{rot} = \int_0^a \frac{2\pi r \, dr}{\pi a^2} \left(\frac{1}{2}M(r\omega)^2\right) = \frac{1}{4}Mv^2.$$

So the total energy is $\frac{1}{2}Mv^2 + \frac{1}{4}Mv^2 = \boxed{\tfrac{3}{4}Mv^2}$.

Parts (b) and (c) are quicker if you know about moments of inertia: The angular momentum is $I\omega$ and the rotational energy is $\frac{1}{2}I\omega^2$, where $I = \int r^2 dM$, the moment of inertia of the cylinder about its axis, is $\frac{1}{2}Ma^2$. But you don't need to know this - you can do the integrals instead!

(d) For a sphere the answer to (a) is unchanged. For (b) and (c) we divide the sphere up into cylindrical shells and proceed as above. We have

$$dM = \frac{2\pi r dr \times 2(a^2 - r^2)^{1/2}}{\frac{4}{3}\pi a^3}M,$$

and

$$L = \int_0^a \omega r^2 dM = \frac{3M\omega}{a^3}\int_0^a r^3(a^2 - r^2)^{1/2}dr = \boxed{\frac{2}{5}M\omega a^2},$$

$$E_{rot} = \frac{3M\omega^2}{2a^3}\int_0^a r^3(a^2 - r^2)^{1/2}dr = \frac{1}{5}M\omega^2 a^2,$$

or $\boxed{E_{tot} = \frac{7}{10}Mv^2}$. More easily, the moment of inertia of a sphere is $I = \frac{2}{5}Ma^2$
so $L = I\omega = \frac{2}{5}Ma^2(v/a)$, and $E_{rot} = \frac{1}{2}I\omega^2 = \frac{1}{5}Mv^2$.

(e) If the disc is slipping then, for a given total energy, its rotational energy and angular momentum will be lower and its translational energy and linear momentum higher.

10.2 Galileo's legendary experiment

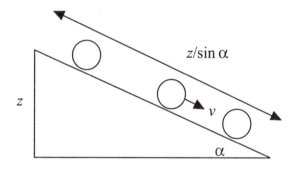

Figure 31. Rolling down an inclined plane

Let the object rolling down the plane have mass M, moment of inertia I, radius a, and a speed v after it has descended a vertical height z. We shall obtain the equation of motion by differentiating the energy equation.

Conservation of energy (K.E. + P.E. = constant) gives

$$\frac{1}{2}M\left(1 + \frac{I}{Ma^2}\right)v^2 - Mgz = \text{constant}. \tag{17}$$

Differentiating equation (17) with respect to time gives

$$\left(1 + \frac{I}{Ma^2}\right)v\dot{v} - g\dot{z} = 0,$$

where $\dot{z} = v \sin \alpha$. Solving for the acceleration $\dot{v}$ gives
$\dot{v} = g \sin \alpha / (1 + I/Ma^2)$, which is constant. We can therefore use the constant acceleration formulae to derive the motion. Using $s = at^2/2$ with $a = \dot{v}$ gives

$$t = \left[\frac{2z \left(1 + \frac{I}{Ma^2}\right)}{g \sin^2 \alpha} \right]^{1/2}.$$

Thus all bodies with the same value of I/Ma^2 fall the same distance in the same time. For any sphere of uniform density $I/Ma^2 = 2/5$ so (a) the composition of the balls has no effect and (b) the size of the balls has no effect. But (c) if we change the shape then I/Ma^2 can change and the times will differ. For example, if Galileo had used a sphere and a cylinder the times would differ by a factor $(14/15)^{1/2}$.

(d) In principle therefore, Galileo could not have verified the universality of free fall (that *all* bodies fall with the same acceleration under gravity) by this method! However, since Galileo used his pulse to measure time any difference of this size may have been disguised by the experimental error.

Note that the universality of free fall is incorporated into dynamics by the use of the same mass M to represent two different concepts in equation (22). In Mgz the M represents the coupling between gravity and the body. This is the gravitational mass M_g. In the kinetic energy term M represents the resistance (inertia) of the body to a force; this is the inertial mass M_i. Thus, the universality of free fall is equivalent to the equality $M_g = M_i = M$. The universality of free fall is the postulate on which the general theory of relativity (Einstein's theory of gravity) is built.

10.3   Snooker

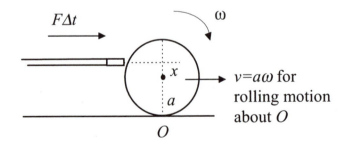

Figure 32. Impact of cue and ball

(a) Let the cue exert an impulse $F\Delta t$ at x from the centre as shown in the figure.

Then conservation of momentum implies:
$$F\Delta t = mv$$
and conservation of angular momentum implies
$$F\Delta t(a + x) = I\omega,$$
where m is the mass of the ball and I is its moment of inertia about O see figure (32). Eliminating F gives
$$(a + x)mv = Iw. \qquad (18)$$
The moment of inertia of the ball (assuming it is of uniform density) about its centre is $I_C = \frac{2}{5}ma^2$, so about O, by the parallel axes theorem, it is,
$$I = \frac{2}{5}ma^2 + ma^2 = \frac{7}{5}ma^2.$$
Substituting for I in (18) gives
$$(a + x)mv = \frac{7}{5}ma^2\omega.$$
For the ball to roll from the start $v = a\omega$, so
$$(a + x) = \frac{7}{5}a,$$
or
$$\frac{(a + x)}{2a} = \frac{7}{10}.$$
Thus the impact point must be $7/10^{\text{ths}}$ the height of the ball above the level of the table.

(b) For impact with the side cushions the same theory applies. Therefore these should be $\boxed{7/10^{\text{ths}}}$ the height of the ball above the table.

(c) A ball not struck at the $7/10^{\text{ths}}$ height (for example, one struck by another ball) will initially slide until friction turns the slide into a roll. As gravity is less on the Moon friction would be lower, so the sliding phase would be longer.

(d) Note that the value of x depends on the moment of inertia I, so a change in I leads to a different x. This means the cushions would be at the wrong height. Furthermore, a gentle shot hit at $7/10^{\text{ths}}$ height would skid and loose speed, probably failing to pot the ball at which it was aimed.

Tutorial 11   Gravitating bodies

11.1   The great escape

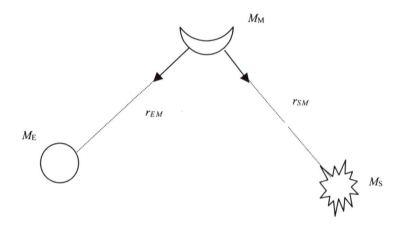

Figure 33. Forces in the Sun-Earth-Moon system

The ratio of the forces on the Moon is

$$\frac{F_S}{F_E} = \frac{M_S}{M_E}\left(\frac{r_{EM}}{r_{SM}}\right)^2$$

$$= \left(\frac{2.0 \times 10^{30} \text{ kg}}{6.0 \times 10^{24} \text{ kg}}\right) \times \left(\frac{3.8 \times 10^5 \text{ km}}{1.5 \times 10^8 \text{ km}}\right)^2$$

$$= 2.1.$$

So the Sun's pull on the Moon is greater than that of the Earth. But the Earth and Moon are both in free fall towards the Sun, so in the Earth-Moon frame the only effect of the Sun is a small tidal force.

11.2   Solar System centre

At some stage all three largest planets will align on the same side of the Sun. Let the centre of mass of these planets and the Sun be a distance x km from the centre of the Sun. Since x will turn out to be small compared to the planetary distances, we can neglect the difference between the distance of a planet from the Sun and its distance from the centre of mass of the system. Taking moments about the centre of mass then gives

$$2.0 \times 10^{30}x \;=\; 1.9 \times 10^{27} \times 7.8 \times 10^8 + 5.7 \times 10^{26} \times 1.4 \times 10^9 + $$

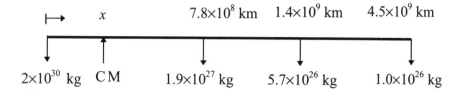

Figure 34. Masses and distances of the major planets

$$+1.0 \times 10^{26} \times 4.5 \times 10^{9}$$
$$= 2.8 \times 10^{36}$$

from which $x = 1.4 \times 10^{6}$ km. This is larger than the radius of the Sun $(0.7 \times 10^{6}$ km).

11.3   Counter-Earth

We assume that both the Earth and counter-Earth are in a circular orbit of radius r. Then balancing the force and central acceleration on the Earth gives

$$\frac{M_E v^2}{r} = \frac{GM_S M_E}{r^2} + \frac{GM_E^2}{(2r)^2}$$

or

$$v^2 = \frac{GM_S}{r}\left(1 + \frac{M_E}{4M_S}\right).$$

So the period is

$$T = \frac{2\pi r}{v} = \frac{2\pi r^{3/2}}{\sqrt{GM_S}}\left(1 + \frac{M_E}{4M_S}\right)^{-1/2}$$

$$\approx T_0\left(1 - \frac{M_E}{8M_S}\right)$$

Here T_0 is the period in the absence of the counter-Earth (obtained by putting the mass of the counter-Earth to zero in the expression for T). We define the difference in orbital period

$$\Delta T = T_0 - T = \frac{M_E}{8M_S}T_0$$

$$= \frac{6 \times 10^{24} \text{ kg}}{8 \times 2 \times 10^{30} \text{ kg}} \times 1 \text{ yr} = \frac{3}{8} \times 10^{-6} \text{ yr or } \boxed{11.9 \text{ s.}}$$

(b) This is only a small effect. However, the arrangement is (marginally) unstable! Suppose the Earth and counter-Earth were to get slightly out of line. Then the force of attraction between them would be as shown in the figure. There is a component of force opposing the motion of the Earth and one in the direction

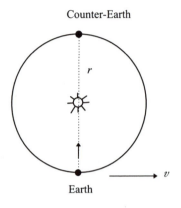

Figure 35. Counter-Earth

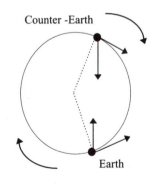

Figure 36. Instability of the counter-Earth

of motion of the counter-Earth. Therefore the Earth will fall to a smaller orbit and speed up and the counter-Earth will move to a more distant orbit and slow down. This will re-align the two bodies with the Sun, but in new orbits. Random perturbations will therefore eventually drive one or the other of the planets into the Sun.

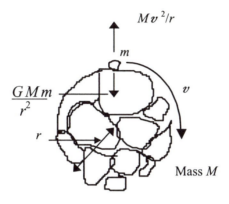

Figure 37. Rubble piles

11.4 Rubble piles

If asteroids are flying rubble piles then the maximum speed of rotation is such that a lump at the surface of the pile must just be in orbit: if the asteroid is assumed to be roughly spherical this condition is

$$\frac{mv^2}{r} = \frac{GMm}{r^2}.$$

The minimum period corresponding to this maximum speed is

$$T = \frac{2\pi r}{v} = \frac{2\pi r^{3/2}}{\sqrt{GM}}$$

Therefore

$$T \geqslant \frac{2\pi r^{3/2}}{\sqrt{GM}} = \frac{2\pi r^{3/2}}{(G \times \frac{4}{3}\pi\rho r^3)^{1/2}} = \frac{2\pi}{(\frac{4}{3}\pi G\rho)^{1/2}},$$

so

$$T \geqslant \frac{2\pi}{[\frac{4}{3}\pi \times (6.7 \times 10^{-11} \text{Nm}^2\text{kg}^{-2}) \times (2000 \text{ kg m}^{-3})]^{1/2}}$$

$$= 8.4 \times 10^3 \text{ s or } \boxed{2.3 \text{ hours.}}$$

11.5   Shuttle dust

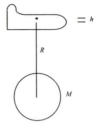

Figure 38. The Space Shuttle in orbit

The centre of mass of the Shuttle in a circular orbit of radius R has an angular velocity w given by

$$Rw^2 = \frac{GM}{R^2}$$

where M is the mass of the Earth. The mote of dust at a distance $R+h$ from the centre of the Earth has a speed $w(R+h)$ which is a little too large for a circular orbit of this radius. It is therefore subject to a net outward acceleration

$$\ddot{h} = (R+h)w^2 - \frac{GM}{(R+h)^2}$$

$$= \frac{GM}{R^3}(R+h) - \frac{GM}{(R+h)^2}.$$

For $h/R \ll 1$ this is

$$\ddot{h} \simeq \frac{GM}{R^3}\left[(R+h) - R\left(1 - \frac{2h}{R}\right)\right]$$

$$= \frac{3GMh}{R^3} \simeq 3g\frac{h}{R}.$$

Solving for h gives

$$h = h_0 e^{(3g/R)^{1/2}t}.$$

So for a Shuttle orbiting at an altitude of 300 km, it takes a time

$$t = \frac{\log_e\left(\frac{h}{h_0}\right)}{\left(\frac{3g}{R}\right)^{1/2}} = \frac{\log_e 2}{\left(\frac{3\times9.81}{6.67\times10^6}\right)^{1/2}} = \boxed{330 \text{ s}}$$

for the dust to move from an initial displacement of 1m from the C.M. to 2m.

Tutorial 12   Oscillations

12.1   Journey to the centre of the Earth

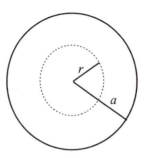

Figure 39. The force at a radius r depends on the mass only within r

(a) At distance r from the centre of the (spherical) Earth of radius a your weight depends on the mass within r as in figure (39). So, assuming an Earth of uniform density, and mass M_E,

$$mg(r) = \frac{GM(r)m}{r^2} = \frac{GM_E m}{a^2}\frac{M(r)}{M_E}\frac{a^2}{r^2} = gm\frac{r^3}{a^3}\frac{a^2}{r^2} = gm\frac{r}{a}.$$

where g is the acceleration due to gravity at the Earth's surface. The equation of motion of a body falling at radius r is therefore

$$m\ddot{r} = -mg\frac{r}{a} \quad \text{or} \quad \ddot{r} = -g\frac{r}{a}.$$

This is simple harmonic motion.

(b)The oscillation frequency is ω given by

$$\omega^2 = \left(\frac{2\pi}{T}\right)^2 = \frac{g}{a} = \frac{GM_E}{a^3}$$

Therefore the period is

$$T = \frac{2\pi a^{3/2}}{(GM_E)^{1/2}} = \frac{2\pi \times (6370 \times 10^3 \text{ m})^{3/2}}{(6.67 \times 10^{-11} \text{ Nm}^2\text{kg}^{-2})^{1/2} \times (6 \times 10^{24} \text{ kg})^{1/2}} = 5.05 \times 10^3 \text{ s}.$$

or about 1.4 hours. The journey time is $T/2 = \boxed{42 \text{ minutes}}$.

(c) The vehicle would be in free fall throughout the journey, since only gravitational forces are acting, so the passengers would be weightless.

(d) The radial position of the vehicle follows simple harmonic motion with amplitude a and angular frequency ω.   Thus $r = a\cos(\omega t + \phi)$ and

$\dot{r} = wa \cos(\omega t + \phi)$. The peak velocity is therefore

$$wa = \frac{2\pi}{(5.05 \times 10^3 \text{ s})} \times (6.370 \times 10^6 \text{ m}) = \boxed{7.9 \text{ km s}^{-1}.}$$

(e) The force on the vehicle at radius r towards the mid-point of the shaft is

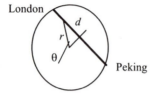

Figure 40. A tunnel between London and Peking

$$F = -\frac{GM(r)}{r^2} \sin\theta = -\frac{GM_E}{a^3} d,$$

where $d = r\sin\theta$ is the distance from the mid-point to the current position of the vehicle. Since the force is again proportional to the displacement this is simple harmonic motion. The period, and hence the time of travel, is the same as for a tunnel through the centre. The maximum speed will be lower, as the amplitude is lower.

12.2   Thomson's current bun

(a) The electrostatic force follows an inverse square law so the force on the displaced electron at radius r comes from the positive charge $Q(r)$ inside a sphere of radius r. If σ is the charge density of positive charge, the restoring force on the electron, charge $-e$, is

$$F = \frac{-Q(r)e}{4\pi\varepsilon_0 r^2} = \frac{-\frac{4}{3}\pi r^3 \sigma e}{4\pi\varepsilon_0 r^2} = \frac{-(\frac{4}{3}\pi a^3 \sigma)er^3}{4\pi\varepsilon_0 r^2 a^3} = \frac{-e^2 r}{4\pi\varepsilon_0 a^3}.$$

Thus, the motion of the electron is exactly simple harmonic. The frequency is

$$f = \frac{\omega}{2\pi} = \left(\frac{e^2}{4\pi\varepsilon_0 a^3 m}\right)^{1/2} \frac{1}{2\pi}$$

$$= \left(\frac{(1.60 \times 10^{-19} \text{ C})^2}{4\pi \times (8.85 \times 10^{-12} \text{ F m}^{-1}) \times (10^{-10} \text{ m})^3 \times (9.1 \times 10^{-31} \text{ kg})}\right)^{1/2} \times \frac{1}{2\pi}$$

$$= \boxed{2.5 \times 10^{15} \text{ Hz}.}$$

(b) On this picture the hydrogen atom would have a single frequency since the frequency of vibration of the electron is independent of amplitude. The actual

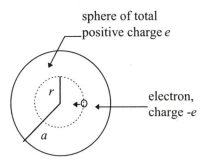

Figure 41. A displaced electron in a sphere of positive charge

spectrum consists of a number of series of the form

$$f = Rc \left(\frac{1}{n^2} - \frac{1}{m^2} \right),$$

with n and $m > n$ integers.

12.3   Satellite attitude

Let the satellite be inclined at an angle θ to the radius vector as shown in figure 42. The upper mass is moving at a radius of $r + l \cos \theta$. The lower mass is at a radius $r - l \cos \theta$. Let $\Delta r = l \cos \theta$ and let the angular velocity of the system be ω. We compute the forces on the system in this configuration. In the rest frame of the satellite, the net outward force on the upper mass is

$$m(r + \Delta r)\omega^2 - \frac{GMm}{(r + \Delta r)^2} \simeq m\omega^2 r(1 + \frac{\Delta r}{r}) - \frac{GMm}{r^2}\left(1 - \frac{2\Delta r}{r}\right)$$

since $\Delta r / r \ll 1$. But, for the orbit of the centre of mass

$$r\omega^2 = \frac{GM}{r^2}.$$

Therefore, the net force is

$$m\omega^2 \Delta r + 2m\omega^2 \Delta r = 3m\omega^2 \Delta r.$$

For small Δr this is also the net inward force on the lower mass. Therefore the *restoring* torque on the system is

$$\tau = 2 \times (3m\omega^2 \Delta r) \times l \sin \theta$$

which gives the equation of motion for small θ

$$\ddot{\theta} = -\frac{6m\omega^2 l^2 \theta}{I} = -3\omega^2 \theta$$

where $\Delta r = l \cos \theta \simeq l$ has been used and the moment of inertia $I = 2ml^2$.

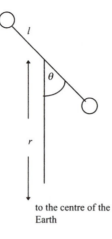

to the centre of the
Earth

Figure 42. A dumbbell satellite

(b) The period of oscillation of the satellite is
$$T = \frac{2\pi}{\sqrt{3}\omega}$$
where $\omega = (GM/r^3)^{1/2}$ is the angular frequency of the orbital motion. The period is therefore
$$T = \frac{2\pi}{\sqrt{3}} \times \frac{r^{3/2}}{(GM)^{1/2}}.$$
This is independent of the mass and length of the satellite.

(c) Let a general position of the satellite be described by the polar angles θ and ϕ relating a frame fixed in the satellite with its axis along the z-direction and a frame fixed in the Earth with the radial vector joining the centre of the satellite to the centre of the Earth along the z-direction. There is no restoring couple to prevent the satellite spinning about its axis (ϕ =constant), but the restoring force governing the θ-motion is independent of ϕ for small displacements in θ. The satellite can therefore spin about the satellite-Earth line.

12.4   Non-harmonic forces

(a) The potential is plotted in figure 43 for values of a and r_0 corresponding approximately to the data given in part (e).

(b) If the interatomic force were exactly harmonic the mean displacement of an atom from its equilibrium position would be zero, independent of temperature. But it is easier to expand a solid than to compress it, so the potential describing the average energy per pair of atoms must be asymmetric with respect to increase

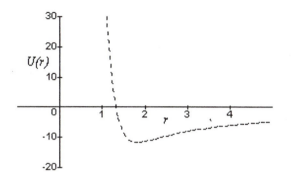

Figure 43. The potential $-24/x + 0.6 \exp(10 - 5x)$

and decrease of the mean separation. A particle oscillating in an asymmetric potential spends more time on the shallower side (weaker restoring force) than the steeper side (stronger restoring force). Therefore a particle oscillating in a potential that is asymmetric about equilibrium acquires a non-zero average displacement that depends on energy. For a potential of the given form, the average displacement is to larger separation and hence to thermal expansion.

(c) We have
$$U_0 = U(r_0) = a/r_0 - (\rho_0 a/r_0^2) = 0.9a/r_0$$
and
$$U_0' = \left[\frac{dU}{dr}\right]_{r=r_0} = -\left[\frac{a}{r^2} - \frac{a}{r_0^2}e^{-(r-r_0)/\rho_0}\right]_{r=r_0} = 0$$
which confirms that $r = r_0$ is the equilibrium separation. Furthermore
$$U_0'' = [d^2U/dr^2]_{r=r_0} = -2\frac{a}{r_0^3} + \frac{a}{r_0^2\rho_0} \simeq 9U_0/r_0^2$$
and,
$$U_0''' = [d^3U/dr^3]_{r=r_0} = 6\frac{a}{r4} - \frac{a}{r_0^2\rho_0^2} \simeq -100U_0/r_0^3$$
The Taylor expansion of $U(r - r_0)$ is therefore
$$U(r - r_0) = U_0(1 + \frac{1}{2}\frac{9}{r_0^2}(r - r_0)^2 - \frac{1}{6}\frac{100}{r_0^3}(r - r_0)^3 + \ldots).$$
(d) For small oscillations $x = r - r_0$ about the equilibrium position the equation of motion is derived from the constancy of the total energy
$$\dot{E} = \frac{d}{dt}(\frac{1}{2}m\dot{x}^2 + \frac{1}{2}U_0''x^2) = 0.$$
We get the standard result $m\ddot{x} = -U_0''x$.

(e) The oscillation frequency is given by $w^2 = U_0''/m = 9U_0/r_0^2 m$. If $U_0 = 6$ eV, $m = \mu m_p$ and $r_0 = 2$Å we get

$$w^2 = \frac{9 \times (6 \times 1.6 \times 10^{-19} \text{ J eV}^{-1})}{(2 \times 10^{-10} \text{ m})^2 \times (1.6 \times 10^{-27} \text{ kg})\mu}$$

or $w \approx \boxed{3.5 \times 10^{14}\mu^{-1/2} \text{ Hz}}$. This is the maximum phonon frequency. (The atoms vibrate independently, so the wavelength is about an atomic spacing; other modes in which the atoms vibrate together have longer wavelengths and hence lower frequencies.)

The Taylor expansion can now be written

$$U \sim U_0 + \frac{1}{2}mw^2(r - r_0)^2 - \gamma(mw^2/r_0)(r - r_0)^3$$

where we have written γ for the factor $\sim 100/6$ in the third term. The equation of motion is again obtained by differentiating the total energy

$$\dot{E} = \frac{d}{dt}(\frac{1}{2}m\dot{x}^2 - U) = 0$$

to get

$$\ddot{x} = -w^2 x + \gamma(w^2/r_0)x^2$$

neglecting factors of order unity.

(f) For small amplitude oscillations, $x/r \ll 1$, the final term in the equation of motion is small. A zeroth order approximate solution is obtained by neglecting it completely, giving $x = A\sin(wt + \phi)$. The phase of the oscillation ϕ is not important so we can take $\phi = 0$ for simplicity. The mean energy of a harmonic oscillator is $mw^2 A^2$ which must equal kT for an oscillator in thermal equilibrium at temperature T. Therefore we get $A^2 = kT/mw^2$.

Now the next approximation to the equation of motion is obtained by using the zeroth order solution to estimate the small term in x^2. This gives

$$\begin{aligned} \ddot{x} &= -w^2 x + \gamma(w^2 A^2/r_0)\sin^2 wt \\ &= -w^2 x + \gamma(w^2 A^2/2r_0)(1 - 2\cos 2wt) \end{aligned} \tag{19}$$

from which

$$x = \gamma A^2/2r_0 + a\cos 2wt + b\sin 2wt + A\sin wt$$

(g) The average dispacement is $\bar{x} = \gamma A^2/2r_0 \sim \gamma kT/(2mw^2 r_0)$. This is the additional mean separation δl of the atoms as a result of their thermal motion. The fractional change in separation is

$$\frac{\delta l}{l} = \frac{\bar{x}}{r_0} = \frac{\gamma kT}{2mw^2 r_0^2}$$

$$= \frac{(100/6) \times (1.4 \times 10^{-23} \text{ J K}^{-1})}{2 \times \mu \times (1.6 \times 10^{-27} \text{ kg}) \times (3.5 \times 10^{14}\mu^{-1/2} \text{ Hz})^2 \times (2 \times 10^{-10} \text{ m})^2}T$$

$$= 1.4 \times 10^{-5}T.$$

The coefficient of thermal expansion is therefore $\boxed{1.4 \times 10^{-5}\,\mathrm{K}^{-1}}$ and is independent of the atomic mass.

We can understand the form of the coefficient of expansion as follows: We must have $\delta l \propto kT$ and the coefficient of proportionality must be a length/energy constructed from the second and third terms of the Taylor expansion of U, hence from m, w and r_0. This gives

$$\frac{\delta l}{r_0} \sim \frac{1}{r_0} \times \frac{(mw^2/r_0)}{(m^2w^4)}kT = \frac{kT}{mw^2r_0^2}$$

as above.

(h) The coefficient of expansion is of the right order of magnitude for a simple ionic solid. Since the pair potential is generally regarded as a rather poor model in many respects, it is interesting that it gives this coefficient to about the right order of magnitude without any further adjustable parameters. This does not imply that it can be used to derive the other anisotropic elastic moduli or even that solids always expand on heating.

Tutorial 13   Resonance

13.1   The opera singer and the wine glass

(a) The brittleness of glass arises from a large number of small internal cracks which generate large stresses in regions of high curvature at the tips of the cracks. When supplied with sufficient energy to create new surfaces the cracks can grow, leading to fracture of the material, see tutorial 26.

(b) Wine glasses tend to be made of high quality glass with smaller internal resistance and a high Q value. This enhances the resonance peak.

(c) A small resistance implies a large time constant since $\tau = m/r$. Thus high Q systems take a long time to come into equilibrium.

(d) The note must be close to the resonant frequency of the glass. Since we have chosen a glass with a narrow resonance peak, as a consequence of requiring a high Q, it is important to hit the right note.

(e) We get the power from $P = F_0^2/r$ so we need F_0, and r. The pressure p at the glass is given in terms of the sound intensity by

$$p = 2 \times 10^{-5} \times 10^{L_p/20}\,\mathrm{N\,m^{-2}}.$$

If the singer produces 100 dB of sound at the glass the pressure is $2 \times 10^{-5} \times 10^5 = 2\,\mathrm{N\,m^{-2}}$.

The driving force F_0 on the glass of cross-sectional area A is therefore $2\,\text{N}\,\text{m}^{-2} \times A$. For $A = 100\,\text{cm}^2$ this is

$$F_0 = 2 \times 10^{-2}\,\text{N}.$$

To get r we guess the damping time constant to be around 5 s (for our own not very high quality glasses) and take the mass of the glass, omitting the base and stem which do not resonate, to be about 40 g. Therefore $r = m/\tau = 0.04/5\,\text{kg}$ $\text{s}^{-1} = 0.008$. The sound power at the glass is therefore

$$P = \frac{F_0^2}{2r} = \frac{(2 \times 10^{-2}\,\text{N})^2}{2 \times (0.008\,\text{kg}\,\text{s}^{-1})} = \boxed{0.25\,\text{W}}.$$

(ii) The energy stored is

$$E = \tau P \sim 5\,\text{s} \times 0.25\,\text{W} = \boxed{1.25\,\text{J}}.$$

Now drop the wine glass from 0.5 m. This transfers a potential energy $mgh = 0.05\,\text{kg} \times 10\,\text{m}\,\text{s}^{-2} \times 0.5\,\text{m} = 0.25\,\text{J}$. This is very approximate, but it appears possible that sufficient energy can be stored to break the glass, as is confirmed by experiment.

13.2 The walls of Jericho

A highly resonant structure can be destroyed by a resonant force applied for long enough (rioters overturning cars, for example). A mud brick wall however is not a resonant structure, so, in contrast to the wine glass of the previous question, there is no build up of stored energy. The effect is more likely to have been psychological and the bringing down of the walls a metaphor!

13.3 Keep on swinging

(a) Resolving along the string the equation of motion of the bob is

$$T - mg \cos\theta = ml\dot\theta^2$$

which, with a slight re-arrangement gives the required result.

To obtain an expression for the time dependence of T we introduce the solution of the equation of motion for θ. For small displacements, resolving horizontally, we have

$$\ddot\theta = -\frac{g}{l}\theta$$

and hence $\theta = \theta_0 \cos(t\sqrt{g/l})$. Therefore

$$
\begin{aligned}
T &\simeq mg\left(1 - \tfrac{1}{2}\theta^2\right) + ml\dot\theta^2 \\
&= mg\left(1 - \tfrac{1}{2}\theta_0^2 \cos^2(t\sqrt{g/l}) + \theta_0^2 \sin^2(t\sqrt{g/l})\right)
\end{aligned}
$$

$$= mg\left(1 + \theta_0^2 - \tfrac{3}{2}\theta_0^2\cos^2(t\sqrt{g/l})\right).$$

Using $\cos^2\theta = (1 + \cos 2\theta)/2$ gives

$$T = mg\left(1 + \tfrac{1}{4}\theta_0^2 - \tfrac{3}{4}\theta_0^2\cos(2t\sqrt{g/l})\right) \tag{20}$$

as required.

(b) We now assume that the additional vertical motion of the support causes only a small perturbation to the swing of the pendulum. Thus, to a sufficient approximation, the force on the support is given by (20) and the support moves with speed

$$v = \frac{d}{dt}(-a\sin 2\omega t) = -2a\omega\cos 2\omega t.$$

where $\omega = \sqrt{g/l}$. The power P delivered equals the rate of doing work, so

$$P = Tv = mg\left(1 + \tfrac{1}{4}\theta_0^2 - \tfrac{3}{4}\theta_0^2\cos(2\omega t)\right)(-2a\omega\cos 2\omega t).$$

To obtain the mean power we average over a cycle. The first two terms do not contribute since the average of $\cos\theta$ over a period is zero. In the final term the average of $\cos^2 2\omega t$ gives a factor of $1/2$. Explicitly, the final term contributes

$$\langle P\rangle = \tfrac{3}{2}\theta_0^2 mga\omega\frac{\int_0^{2\pi/\omega}\cos^2 2\omega t\, dt}{2\pi/\omega}$$

$$= \boxed{\tfrac{3}{4}\theta_0^2 mga\omega.}$$

Note that it is the component of motion of the support in phase with the motion of the pendulum that delivers energy. However, contrary to what one might guess the support needs to be moved at twice the frequency of the pendulum. Physically, the raising of the support every time the bob passes through the origin feeds potential energy into the system.

(c) The energy stored in the pendulum is $E = \tfrac{1}{2}ml^2\langle\dot\theta^2\rangle + \tfrac{1}{2}m\omega^2 l^2\langle\theta^2\rangle = \tfrac{1}{2}mgl\theta_0^2$. But

$$\dot E = \langle P\rangle = \tfrac{3}{4}\theta_0^2 mga\omega = \frac{3}{2}\frac{a\omega}{l}\times\frac{1}{2}mgl\theta_0^2 = \frac{E}{\tau}$$

where $\tau = 2l/(3a\omega)$. Thus the energy grows exponentially on a timescale τ.

In practice the pendulum at the London Science Museum and the child's swing suffer damping so the amplitude of oscillation builds up until the dissipated power equals the input power.

13.4 Coupled oscillators

(a) Define $X = x_1 + x_2$ and $Y = x_1 - x_2$. Then by adding the equations of motion we get

$$\ddot{X} = -(\omega^2 - g)X$$

and by subtracting

$$\ddot{Y} = -(\omega^2 + g)Y.$$

These are simple harmonic motion at the two slightly different frequencies $\omega_1 = \sqrt{\omega^2 - g} \simeq \omega - g/2\omega$ and $\omega_2 \simeq \omega + g/2\omega$. The effect of the coupling is therefore to split the resonant frequency of the subsystems into two. The initial conditions at $t = 0$ are $X(0) = x_1(0) + x_2(0) = 2a$, $\dot{X}(0) = 0$, $Y(0) = x_1(0) - x_2(0) = 2a$, $\dot{Y}(0) = 0$. Hence

$$\begin{aligned} X &= 2a \cos \omega_1 t \\ Y &= 2a \cos \omega_2 t. \end{aligned}$$

Solving for x_1 and x_2 we find

$$\begin{aligned} x_1 &= a(\cos \omega_1 t + \cos \omega_2 t) = 2a \cos(gt/2\omega) \cos \omega t \\ x_2 &= a(\cos \omega_1 t - \cos \omega_2 t) = 2a \sin(gt/2\omega) \sin \omega t. \end{aligned}$$

The motion of each oscillator is therefore made up of the sum of two simple harmonic motions at the two slightly different natural frequencies of the coupled system, but with different phases. At the start the second oscillator is at rest and the first in motion. The amplitude of motion of the second builds up on a timescale $2\omega/g$ while that of the first goes down on the same timescale. Then the process is reversed. Energy is therefore transferred between the two oscillators at the beat frequency. In practice, of course, dissipation will prevent this process from continuing indefinitely.

(b) Energy will be transferred principally from pendulum C to the pendulum with the same (or nearest) natural frequency, that is to the pendulum which is closest to it in length. Other pendulums of differing lengths will pick up some motion depending on the width of their resónance curves.

(c) If the atom is isolated from electromagnetic fields then these field oscillators must be considered classically to have zero energy. The situation is similar to that of part (b). Accordingly the atomic oscillator would share a fraction of its energy with all the field oscillators to which it is coupled. But in the present case there are an infinite number of coupled field oscillators in any band of frequency, so these provide an infinite sink of energy for the atomic oscillator. Therefore the oscillator, and hence the motion of the electrons which it represents would be completely damped. This is another version of the argument that the classical planetary model of the atom cannot be stable.

(d) In the quantum theory there will still be a transfer of energy from an atomic oscillator to the field, but there can now also be a transfer of the zero point energy of the field oscillators back to the atom. The net result is that there is a lowest energy state of the atomic oscillator (of non-zero energy) in which the damping effect of the field is exactly balanced by its driving effect. This is the stable ground state of the atom. (see Senitsky, 1961, *Phys Rev* **124**, 642).

Tutorial 14   Antigravity

14.1   Cavorite

(a) Let the particles have masses $-m_1$ and $-m_2$ and separation r. Newton's law of gravity applied to particle 1 gives

$$F_1 = G\frac{(-m_1)(-m_2)}{r^2} = G\frac{m_1 m_2}{r^2}$$

towards particle 2, and a force of the same magnitude on particle 2 acting in the direction of particle 1. So the force on each particle acts towards the other. *How-*

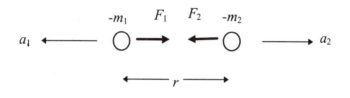

Figure 44. Particles of negative mass repel

ever, Newton's second law gives the acceleration of each particle in response to this force as

$$F_1 = -m_1 a_1; \qquad F_2 = -m_2 a_2.$$

So the acceleration is in the opposite direction to the force for each particle. Thus they repel each other.

(b) If particle 1 now has positive mass the force on particle 1 is

$$F_1 = -G\frac{m_1 m_2}{r^2}$$

acting away from particle 2, with a force on particle 2 of the same magnitude acting away from particle 1.

Applying Newton's second law we see that particle 1 accelerates in the direction of the applied force and particle 2 in the direction opposite to the applied force. Consequently both particles accelerate in the same direction.

This violates Newton's third law (that action and reaction are equal and opposite). If $m_1 = m_2$ (in magnitude) the system acquires no net momentum or energy (because $m_1 v_1 - m_2 v_2 = 0$), but if $m_1 \neq m_2$ momentum is not conserved.

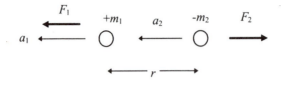

Figure 45. Masses of opposite sign

14.2   Cosmological constancy

(a) A test particle of mass m embedded in a distribution of negative mass is
affected only by the mass $M_{-ve} = \frac{4}{3}\pi r^3 \rho_{-ve}$ within its orbit of radius r. So the
net force acting on it, including that of the Sun, mass M, is

$$\text{Force} = -\frac{GMm}{r^2} - G\frac{\frac{4}{3}\pi r^3 \rho_{-ve} m}{r^2} \tag{21}$$

$$= -\frac{GMm}{r^2} + \frac{1}{3}\Lambda mr$$

where $\Lambda = -4\pi G \rho_{-ve}$ or $\rho_{-ve} = -\Lambda/4\pi G$.

(b) From equation (21) the planetary orbit, assumed circular, is given by

$$\frac{mv^2}{r} = \frac{GmM}{r^2}\left(1 - \frac{\frac{4}{3}\pi |\rho| r^3}{M}\right) = \frac{GmM}{r^2}\left(1 - \frac{M_{-ve}}{M}\right).$$

The orbital period is

$$T = \frac{2\pi r}{v} = \frac{2\pi r^{3/2}}{(GM)^{1/2}\left(1 - \frac{M_{-ve}}{M}\right)^{1/2}} \simeq \frac{2\pi r^{3/2}}{(GM)^{1/2}}\left(1 + \frac{M_{-ve}}{M}\right).$$

If $|\rho| = \Lambda/4\pi G = (10^{-35}\text{ s}^{-2})/4\pi G$, the fractional change in period is

$$\frac{\Delta T}{T} = \frac{\frac{4}{3}\pi r^3 \rho}{M} = \frac{4\pi \times (1.5 \times 10^{11}\text{ m})^3 \times (10^{-35}\text{ s}^{-2})}{3 \times (2 \times 10^{30}\text{ kg}) \times 4\pi \times (6.67 \times 10^{-11}\text{ N m}^2\text{ kg}^{-2})}$$

$$= \boxed{8.4 \times 10^{-23}}$$

which is entirely negligible.

14.3   Bubble gravity

The force from bubble 1 on bubble 2 would be cancelled by introducing a sym-
metrically placed third bubble as shown in the figure. Therefore the force on
bubble 2 is that due to the gravitational field of the mass displaced by bubble 3.
At bubble 2, this mass M generates an acceleration $g = GM/r^2$, where r is the
separation of the bubbles. This generates an upthrust in the negative z direction
on bubble 2 equal to $\rho V g$, where ρ is the density of the fluid, V the volume of

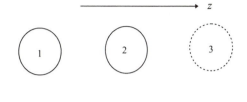

Figure 46. Gravitating bubbles in a uniform fluid

a bubble, which pushes bubble 2 towards bubble 1. By symmetry, bubble 1 also moves towards bubble 2. Thus, the two bubbles *attract* as if they were positive masses acting under Newtonian gravity, rather than repel as would two negative masses.

Tutorial 15   Orbits of the Sun

15.1   The wasting Sun

(a) Mass loss by radiation per year

$$
\begin{aligned}
&= \text{Energy loss per year}/c^2 \\
&= \text{Luminosity} \times (3 \times 10^7 \text{ s yr}^{-1})/c^2 \\
&= \frac{(3.9 \times 10^{26} \text{ W}) \times (3 \times 10^7 \text{ s yr}^{-1})}{(3 \times 10^8 \text{ m s}^{-1})^2} \\
&= 1.3 \times 10^{17} \text{ kg yr}^{-1},
\end{aligned}
$$

or $\boxed{6.8 \times 10^{-14} M_\odot \text{ yr}^{-1}.}$ This is larger than the loss due to the solar wind by a factor of about 3.

(b) The changes in the mass of the Sun causes changes in the orbit of the Earth. Conservation of angular momentum implies $vR = h = \text{constant}$

The acceleration due to gravity equals centripetal acceleration so $v^2/R = GM/R^2$. Hence $MR = v^2 R^2/G = h^2/G = \text{constant}$, as required.

(c) The mass loss over 10^9 years (say) from part (a) is

$$
\begin{aligned}
\Delta M &= (6.8 \times 10^{-14} M_\odot \text{ yr}^{-1}) \times (10^9 \text{ yr}) \\
&= 6.8 \times 10^{-5} M_\odot.
\end{aligned}
$$

Now use $MR = \text{constant}$ so $\Delta R/R = -\Delta M/M$ and hence

$$
\Delta R = 6.8 \times 10^{-5} R = 6.8 \times 10^{-5} \text{ a.u.}
$$

This is a negligible change in the Earth's orbit.

15.2 Orbital impact

The collision sets the Earth off at a new speed determined by conservation of linear momentum. With this new speed a previously circular orbit becomes an ellipse. The properties of the ellipse are determined by conservation of energy and conservation of angular momentum. Since the collision slows the Earth down, the point of collision becomes aphelion. The point of the question is therefore to determine how much closer to the Sun the new perihelion is.

Consider first the initial collision:

Conservation of linear momentum gives: $Mv - mv = (M + m)u$.

This determines the speed after collision:

$$u = \left(\frac{M - m}{M + m} \right) v. \tag{22}$$

As $m/M \ll 1$ this can be written as

$$u \simeq \left(1 - \frac{2m}{M} \right) v.$$

The original orbit determines the initial speed, v, by

$$v^2 = \frac{GM_\odot}{R}.$$

The change in orbit is shown in figure (49). To find the new distance, $r = SP$, in terms of the radius of the previous circular orbit, $R = SA$, we apply conservation of angular momentum and energy to the Earth+asteroid (mass $M + m$) at A and P:

Conservation of angular momentum: yields

$$(M + m)uR = (M + m)Vr;$$

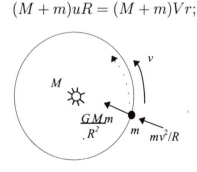

Figure 47. The Earth in orbit

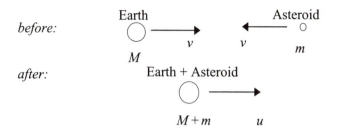

Figure 48. The initial collision

and conserving energy gives

$$\frac{1}{2}u^2 - \frac{GM_\odot}{R} = \frac{1}{2}V^2 - \frac{GM_\odot}{r}.$$

Now we eliminate V to get (after some algebra)

$$r = \frac{u^2 R}{\left(\frac{2GM_\odot}{R} - u^2\right)} = \frac{u^2 R}{2v^2 - u^2}.$$

Since we know u from equation (22) this gives

$$r = \frac{(1 - 4m/M)}{(1 + 4m/M)} R \simeq R\left(1 - \frac{8m}{M}\right).$$

where $(1 - 2m/M)^2 \simeq (1 - 4m/M)$ has been used. Let us put in some numbers. Since the Earth and the asteroid have the same density, the ratio of the masses is $m/M = (r_A/r_E)^3 \simeq 10^{-9}$. So the change in radius of the Earth's orbit is

$$8(m/M)R = (8 \times 10^{-9}) \times (1.5 \times 10^8 \text{ km}) = 1.2 \text{ km}.$$

This is too small to effect a change in the Earth's climate.

15.3 Hale-Bopp spectacular

(a) At escape velocity the comet would have to have just sufficient kinetic energy to overcome gravity, so: $\frac{1}{2}mv_{esc}^2 = GMm/r$, or $v_{esc} = \sqrt{2GM/r}$. At perihelion this is

$$v_{esc} = \left[\frac{2 \times (6.673 \times 10^{-11} \text{ N m}^2 \text{ kg}^{-2}) \times (1.989 \times 10^{30} \text{ kg})}{0.914 \times (1.496 \times 10^{11} \text{ m})}\right]^{1/2}$$

$$= \boxed{44.06 \text{ km s}^{-1}}$$

(b) To determine the orbital parameters we again use conservation of angular momentum and energy. Angular momentum conservation gives

$$vr = VR,$$

and energy conservation
$$\frac{1}{2}mv^2 - \frac{GMm}{r} = \frac{1}{2}mV^2 - \frac{GMm}{R}.$$
We know v and r (or v_{esc}) and we want R, so we eliminate V, and, following the hint in the questions, introduce v_{esc} :
$$v^2 - v_{esc}^2 = \frac{v^2 r^2}{R^2} - \frac{v_{esc}^2 r}{R}$$
or
$$v^2 \left(1 - \frac{r^2}{R^2}\right) = v_{esc}^2 \left(1 - \frac{r}{R}\right).$$
At this point we can divide out $(1 - r/R)$ (instead of solving a quadratic) to get, finally,
$$R = \frac{r}{\frac{v_{esc}^2}{v^2} - 1}. \tag{23}$$
If the observed speed at perihelion was 44.01 km s^{-1} this gives
$$R = \frac{0.914}{\left(\frac{44.06}{44.01}\right)^2 - 1} = \boxed{402 \text{ AU}}$$
(c) For the period we use Kepler's third law $T^2 = \frac{4\pi^2 a^3}{GM}$. If a is in AU, M in solar masses and T in years then this is $T^2 = a^3$ (because 1 year corresponds to 1AU). So
$$T = \left(\frac{r + R}{2}\right)^{3/2} = \left(\frac{0.914 + 402}{2}\right)^{3/2} = 2859 \text{ yr.}$$
The comet will next be at perihelion in the year $1997 + 2859 = \boxed{4856 \text{ AD}}$.

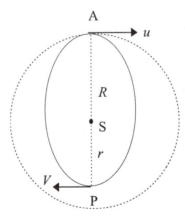

Figure 49. The new orbit

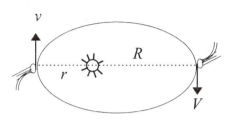

Figure 50. A cometary orbit

(d) This is likely to be inaccurate since $R \propto (v_{esc}^2 - v^2)^{-1}$ (from 23) and $v_{esc} \approx v$. Thus small errors in the speed at perihelion can lead to large errors in R and in T. The best estimate of the period of Hale-Bopp as of February 1998 is 2392 years to within a few years.

We have fudged the issue by quoting the speed at perihelion as if it were an observation rather than (as it is in practice) an inference from the computed orbit based on positional and timing information.

Tutorial 16   Accelerated frames

16.1   Nevertheless it moves

(a) As soon as the stone is thrown into the air it is no longer forced to move horizontally whilst the point on the Earth's surface from which it was projected continues to move eastwards at a speed $R\omega$.

The time in the air is twice the time to reach 10 m which is obtained from $s = \frac{1}{2}gt^2$. So

$$t_{total} = 2 \times \left(\frac{2 \times 10 \text{ m}}{9.81 \text{ m s}^{-2}} \right)^{1/2} = 2.86 \text{ s}$$

(This time is long enough that it could be estimated directly from observation, for example using the pulse to measure time; formulae for constant acceleration were not known until the fourteenth century.) The distance travelled by the stone to the west in this time should be

$$R\omega t = (6370 \text{ km}) \times \left(\frac{2\pi}{24 \times 3600} \text{ rad s}^{-1} \right) \times (2.86 \text{ s}) = \boxed{1.3 \text{ km}}.$$

This is something one could hardly fail to notice, so the conclusion based on Aristotelian dynamics is that the Earth cannot be rotating. (A more contemporary analogue is provided by the Michelson-Morley experiment. In the context of

Newtonian physics this 'proves' that the Earth has no significant orbital motion!)

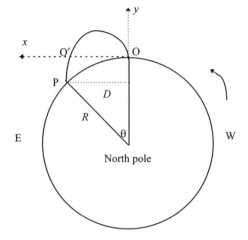

Figure 51. A projectile launched vertically at the equator

(b) Figure 51 shows the trajectory of a projectile launched vertically from the point O on the equator as viewed from an inertial (non-rotating) frame at the north pole. Let the projectile be in flight for a time t. In this time the point O moves through an angle ωt and hence through an arc of length $s = R\omega t$. The projectile inherits the linear motion of the Earth at its point of launch, so has an intial speed along the x-axis of $R\omega$. Suppose it travels a distance D in the x-direction before striking the Earth again. It therefore travels through an arc of length

$$S = R\theta = R\sin^{-1}\frac{D}{R} \simeq R\left(\frac{D}{R} + \frac{1}{6}\frac{D^3}{R^3}\right).$$

The displacement on the ground between the point of launch and the point of impact is

$$\delta = S - s \simeq D + \frac{1}{6}\frac{D^3}{R^2} - R\omega t. \qquad (24)$$

It remains to compute the distance D and the time t. The time to reach P from O' is a relatively small quantity which can be ignored. (An accurate calculation involves a determination of the point of intersection of the trajectory of the projectile with the circle representing the equator.) To obtain the time to reach O' it is sufficient to treat the vertical motion as a projectile under a constant acceleration g. Given that the height reached is 10 m the time is 2.86 s, as determined in part (a). We can therefore put $t = 2.86$ s.

However, for the distance D, we cannot ignore the fact that the direction of the acceleration is changing slightly during the motion of the projectile. This provides a variable component of acceleration in the x-direction, and the trajectory of the projectile departs significantly from a parabola. To compute the distance D we must compute this horizontal motion.

The x-component of the equation of motion is

$$\ddot{x} = -g\sin\theta \simeq -g\theta \simeq -g\frac{x}{R} \tag{25}$$

for small θ. To integrate (25) we use $\ddot{x} = v\, dv/dx$. This gives

$$\int_{R\omega}^{v} v\, dv = \int_{0}^{x} -\frac{gx}{R}\, dx,$$

since the initial speed in the x-direction is $R\omega$, and hence

$$v = \frac{dx}{dt} = \left(R^2\omega^2 - \frac{gx^2}{R}\right)^{1/2} \simeq R\omega\left(1 - \frac{gx^2}{2R^3\omega^2}\right).$$

by the binomial theorem. Integrating again,

$$\int_{0}^{D} \frac{dx}{\left(1 - \frac{gx^2}{2R^3\omega^2}\right)} \simeq \int_{0}^{D}\left(1 + \frac{gx^2}{2R^3\omega^2}\right) dx = \int_{0}^{t} R\omega\, dt$$

or

$$D + \frac{gD^3}{6\omega^2 R^3} = R\omega t.$$

To solve this for D note that a first approximation is $D = R\omega t$ (for constant acceleration) so finally

$$D \simeq R\omega t - \frac{g\omega}{6}t^3.$$

Substituting into (24) gives for the displacement on landing

$$\delta \simeq R\omega t - \frac{g\omega}{6}t^3 + \frac{1}{6}\frac{(R\omega t)^3}{R^2} - R\omega t$$

$$= -\frac{g\omega}{6}t^3 + \frac{R\omega^3 t^3}{6}.$$

The second term is the sole contribution for a constant speed in the x-direction and the first term comes from the component of g along the x-axis. The first term clearly dominates since the acceleration due to gravity g exceeds the centripetal acceleration $R\omega^2$ due to the Earth's rotation. The angular velocity of the Earth is

$$\omega = \frac{2\pi}{24 \times 3600} \text{ rad s}^{-1} = 7.\,27 \times 10^{-5} \text{ rad s}^{-1}.$$

Thus

$$\delta \approx -\frac{1}{6} \times (9.81 \text{ m s}^{-2}) \times (7.\,27 \times 10^{-5} \text{ rad s}^{-1}) \times (2.86 \text{ s})^3$$

$$= \boxed{-2.8 \times 10^{-3} \text{ m.}}$$

Since $\delta < 0$ the projectile lands to the west of the launch point but the distance is too small to be detectable. The first successful demonstration of the rotation of the Earth by dynamical means was that of Foucault's pendulum.

16.2   The latitude effect

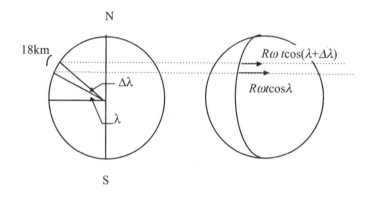

Figure 52. Eastward motion of projectile and target

(a) Let the projectile be launched from a point with latitude λ and land at a latitude $\lambda + \Delta\lambda$ (see figure 52). Let the time of flight of the projectile be t. In this time the target moves with the Earth through an angle wt and hence through an arc of length $s = [R\cos(\lambda + \Delta\lambda)]wt$. The projectile inherits the horizontal velocity of the Earth at the point of launch, $Rw\cos\lambda$. This motion carries it to the east a linear distance $D = twR\cos\lambda$ and hence through an arc on the ground of length $S = R\cos\lambda\sin^{-1}(D/R\cos\lambda) \simeq D$. There is no need to compute S any more accurately: since D/R is of order wt, higher order terms would add corrections of order w^3t^3 which are negligible. Thus, the projectile lands to the east of the target by a distance measured on the Earth of $S - s$

$$= twR\cos\lambda - twR\cos(\lambda + \Delta\lambda)$$

$$\simeq twR\Delta\lambda\sin\lambda \qquad\qquad (26)$$

$$= \left(\frac{18\times10^3 \text{ m}}{300 \text{ ms}^{-1}}\right) \times \left(\frac{2\pi \text{ rad s}^{-1}}{24\times3600}\right) \times (6370 \text{ km}) \times \left(\frac{18\times10^3 \text{m}}{6370 \text{ km}}\right) \times \frac{1}{\sqrt{2}}$$

$$= \boxed{55.5 \text{ m.}}$$

(b) From equation (26) the distance deviated is

$$twR\Delta\lambda\sin\lambda = (6 \text{ s}) \times \left(\frac{2\pi}{24\times3600} \text{ rad s}^{-1}\right) R \times \frac{250 \text{ m}}{R} \times \frac{1}{\sqrt{2}}$$

$$\approx \boxed{8 \text{ cm.}}$$

So the coriolis force is not a problem for golfers.

16.3   The first battle of the Falklands

The navy guns were corrected for the latitude effect in the northern hemisphere. Since $\sin(-\lambda) = -\sin(\lambda)$ in (26) the opposite çorrection is required in the southern hemisphere. Clearly the fleet gunners had not had time to think this out!

16.4   Accelerations of the Earth

The Earth has an orbital acceleration relative to the Sun and it also shares in the Sun's acceleration in the Galaxy, the acceleration of the Galaxy in the Local Group (mainly towards Andromeda) and acceleration of the Local Group in the Local Supercluster (mainly in the direction of Virgo). A true inertial frame would undergo no acceleration with respect to the cosmic background radiation.

Tutorial 17   Artificial gravity

(a) The gravitational field inside an infinite hollow cylinder is zero. To see this we can use Gauss's theorem. If g is the gravitational field (= force per unit mass = acceleration due to gravity) produced by a mass M enclosed by a surface S then

$$\int_S \mathbf{g}.d\mathbf{S} = -4\pi GM$$

If S is a cylindrical surface in the interior of the cylinder $M = 0$ and hence g = 0.

For a finite cylinder there will be edge effects which will give position dependent contributions to g. Instead of trying to estimate these in detail note that the acceleration due to gravity on the inner surface cannot exceed that on the outer surface and must be less for a finite cylinder than for an infinite one. Therefore, we begin by computing the field on the surface of an infinite cylinder. From Gauss's theorem for a cylinder of mass σ per unit length

$$2\pi Rg = 4\pi G\sigma$$

with g in the radial direction. For a cylinder of thickness ΔR of density ρ

$$
\begin{aligned}
g &= \frac{2G \times \pi R \Delta R \rho}{R} = 4G\pi \Delta R \rho \\
&= 4\pi \times (6.67 \times 10^{-11} \text{ N m}^2\text{kg}^{-2}) \times (8500 \text{ kg m}^{-3}) \times \Delta R
\end{aligned}
$$

$$= 7.12 \times 10^{-6} \Delta R \text{ m s}^{-2}.$$

For $\Delta R < R = 10^4$ m we get $g \lesssim 7 \times 10^{-2}$ ms^{-2}. Even this is small, and the true contribution (allowing for the fact that the cylinder is hollow) will be much less.

(b) The effective gravitational acceleration arising from the cylinder's rotation is

$$R\omega^2 = (10 \times 10^3 \text{ m}) \times \left(\frac{2\pi}{4 \times 60} \text{ rad s}^{-1} \right)^2 = \boxed{6.85 \text{ m s}^{-2}.}$$

(c) Let y be the height of the projectile above the inner surface at time t at which point it has travelled a linear distance vt where v is the overall velocity relative to an inertial frame (Figure 53). By the cosine rule

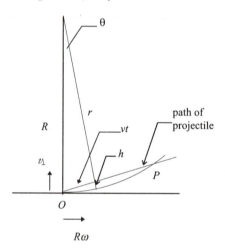

Figure 53. Motion of a projectile launched vertically from the point O viewed from an inertial frame

$$r^2 = R^2 + v^2 t^2 - 2Rvt \cos(\tfrac{\pi}{2} - \theta).$$

Since we are assuming the vertical component of velocity is small compared with the rotational component, the height of the projectile above the ground $h = R - r \ll R$, so

$$r \simeq R - vt \sin\theta + \tfrac{1}{2}\frac{v^2}{R}t^2.$$

Finally, $v^2 = v_\perp^2 + R^2\omega^2 \simeq R^2\omega^2$ and

$$h \simeq v_\perp t - \tfrac{1}{2}gt^2$$

where $g = R\omega^2$ is the effective gravity.

(d) We view the golf shot looking down the axis of the cylinder from an inertial

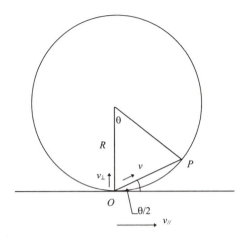

Figure 54. Component of velocity of a projectile viewed along the axis

frame (figure 54). We can ignore the component of velocity of the golf ball along
the axis of the cylinder (the z-axis) and look at the motion in the x-y plane. The
path of the ball is then the line OP. The ball has a component of velocity $v_\perp$ as a
consequence of being struck, and a component $v_\parallel = R\omega$ which it acquires from
the rotation of the cylinder. The path OP is therefore traversed at the constant
speed

$$v = (\omega^2 R^2 + v_\perp^2)^{1/2}$$

in a time t say, and has a length

$$d = R\omega t \left(1 + \frac{v_\perp^2}{R^2\omega^2}\right)^{1/2}.$$

The length of the corresponding arc on the cylinder is $s = 2R\sin^{-1}\frac{d}{2R} \simeq d$ to
sufficient accuracy. In this time the radius vector (and the fairway) turns through
an angle ωt and hence through an arc $s' = R\omega t$. The difference $s - s'$ is displacement of the ball from its line of aim.

$$\delta = s - s' = R\omega t \left(1 + \frac{v_\perp^2}{R^2\omega^2}\right)^{1/2} - R\omega t \simeq \frac{1}{2}\frac{v_\perp^2 t}{R\omega}.$$

If the shot is a drive with velocity 50 ms^{-1} at an angle of 25° with respect to the
ground, then $v_\perp = 50\sin 25° = 21.1$ ms^{-1}. It remains only to calculate the time
of flight. This is determined by the intersection of OP with the cylinder. From
the geometry we have

$$\theta = \tan^{-1}\left(\frac{v_\perp}{R\omega}\right) \simeq \frac{\omega t}{2}.$$

So

$$t = \frac{2}{\omega} \times \tan^{-1}\left(\frac{v_\perp}{R\omega}\right) = \frac{2}{\left(\frac{2\pi}{4\times 60}s^{-1}\right)} \times \tan^{-1}\left(\frac{21.1 \text{ m s}^{-1}}{(10^4\text{m}) \times \left(\frac{2\pi}{4\times 60}s^{-1}\right)}\right) = 6.1 \text{ s}.$$

Finally therefore

$$\delta = \frac{1}{2} \times \frac{(21.1 \text{ m s}^{-1})^2 \times 6}{(10^4 \text{ m}) \times \left(\frac{2\pi}{4\times 60} s^{-1}\right)} \approx \boxed{5 \text{ m}}$$

A deviation of 5m from the line of play is significant so a golfer would have to allow for a mild slice (ball going to the right) when playing 'up' the cylinder and a pull (ball going to the left) when playing in the opposite direction.

(e) The components of motion of the ball along the axis and round the cylinder are independent.   The ball moves along the cylinder without any effect ($\mathbf{v} \wedge \mathbf{\omega} = 0$) and the motion round the cylinder is exactly as it would have been had it not been struck. Therefore, the rotation of Rama has no effect on putting.

(f) Assume the serve is struck parallel to the ground from a height h. It acquires

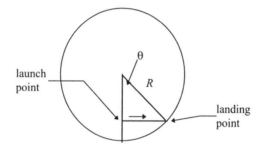

Figure 55. A serve struck horizontally

a transverse component of velocity from the rotation of Rama of $\omega(R - h)$. In a time t it travels a transverse distance $\omega(R - h)t$ and hence through an angle $\theta = \sin^{-1}[\omega(R - h)t/R]$. Thus the ball travels through an arc $R\theta$ while the ground travels through an arc $R\omega t$. The displacement from the line of serve is

$$\begin{aligned}
\delta &= R\theta - R\omega t \\
&= R\sin^{-1}\left[\frac{\omega(R - h)t}{R}\right] - R\omega t \\
&= R\left[\frac{\omega(R - h)t}{R} + \frac{1}{6}\left(\frac{\omega(R - h)t}{R}\right)^3 + ...\right] - R\omega t \\
&\approx -h\omega t + \frac{1}{6}\omega^3 t^3 R.
\end{aligned}$$

For $h = 2$ m and $t = 1$ s this is

$$= -2\,\text{m} \times \left(\frac{2\pi}{4 \times 60}\,\text{rad s}^{-1}\right) \times 1\text{s} + \frac{1}{6}\left(\frac{2\pi}{4 \times 60}\,\text{rad s}^{-1}\right)^3 \times (1\,\text{s})^3 \times (10^4\,\text{m})$$

$$= -5.2 \times 10^{-2} + 3.0 \times 10^{-2}\,\text{m},$$

which is $\boxed{2.2\,\text{cm}}$ and is just about enough for the ball to miss the 'sweet spot' of the racket.

Tutorial 18   Stress and strain

18.1   Towing ropes

(a) For a rope of cross section area A and natural length l_0 stretched to a length l by a force F we have

$$\text{stress} = \frac{F}{A}$$

and

$$\text{strain} = \frac{l - l_0}{l_0}$$

For a Hookean rope stress and strain are related by

$$\text{stress} = Y \times \text{ strain}.$$

The work done is

$$W = \int_{l_0}^{l} F\,dl = \int_{l_0}^{l} \frac{AY(l - l_0)}{l_0}\,dl = \frac{AY(l - l_0)^2}{2l_0} = \frac{1}{2}Y\frac{(l - l_0)^2}{l_0^2}l_0 A, \quad (27)$$

$$= \frac{1}{2}Y \times (\text{strain})^2 \times \text{volume}. \quad (28)$$

This work is stored as strain energy in the rope.

(b) From equation (28) the longer rope can store more energy for a given strain so can soak up larger jerks without breaking.

(c) The kinetic energy of Concorde on landing is

$$\frac{1}{2} \times (10^5\,\text{kg}) \times \left(\frac{2.4 \times 10^2 \times 10^3\,\text{m hr}^{-1}}{3.6 \times 10^3\,\text{s hr}^{-1}}\right)^2 = 2.2 \times 10^8\,\text{J}.$$

The energy input to break a rope is 4×10^7 J m^{-3}, so no more than $(2.2 \times 10^8\,\text{J})/(4 \times 10^7\,\text{J}) = 5.5$ m^3 of rubber is needed. Since the density of rubber is about 1000 kg m^{-3} this amounts to about 5 tonnes of rubber, so the tactic is feasible.

18.2   Jumping flea

(a) For the flea the acceleration is $= 1\text{m s}^{-1}/10^{-3}\text{s} = 10^3\text{ m s}^{-2} \sim 100g$.

The Olympic high jumper gets to a height of 2.4m which raises the centre of gravity by about 1.5 m. Conservation of energy gives

$$\frac{1}{2}mv^2 = mgh \Rightarrow v = (2gh)^{\frac{1}{2}}.$$

So $v = (2 \times 9.81 \times 1.5)^{\frac{1}{2}} = \text{m s}^{-1} = 5.4 \text{ m s}^{-1}$.

The duration of the spring which produces this jump is, say, 0.3 s so the acceleration is 5.4/0.3 m s^{-2} or about $2g$.

(b) The kinetic energy of the flea

$$\text{K.E.} = \frac{1}{2}mv^2 = \frac{1}{2} \times (0.45 \times 10^{-6}\text{ kg}) \times (1\text{ m s}^{-1})^2 = \boxed{2.3 \times 10^{-7}\text{ J}}.$$

(c) The muscle energy = muscle weight × muscle power × time or

$(\frac{20}{100} \times 0.45 \times 10^{-6}\text{ kg}) \times (60\text{ W kg}^{-1}) \times (10^{-3}\text{ s}) = 5.4 \times 10^{-9}\text{ J}$.

So a flea cannot generate the kinetic energy of the jump from muscle power alone.

(d) The energy stored in the pads is $\frac{1}{2}Y \times (\text{strain})^2 \times$ volume. We assume unit strain - larger values are possible in biological materials (but Hooke's law is not then valid) so this is a conservative estimate. The stored energy is then

$$\frac{1}{2}(1.7 \times 10^6\text{ N m}^{-2}) \times 1 \times (1.4 \times 10^{-4} \times 10^{-9}\text{ m}^3) = 1.2 \times 10^{-7}\text{ J}$$

We conclude that the pads can supply the missing energy.

(e) The pads can be compressed on a much longer timescale than 10^{-3} s.

18.3   Into the deep

(a) The pressure P above atmospheric at depth h is

$P = \rho g h$

$= (1025\text{ kg m}^{-3}) \times (9.81\text{ m s}^{-2}) \times (10.9 \times 10^3\text{ m}) = \boxed{1.1 \times 10^8\text{ Pa}}$

where 1 Pascal (Pa) = 1 N m^{-2}. So the pressure is about 1000 atmospheres at the bottom of the trench. (b) From the definition of the bulk modulus B as stress/strain, we have

$$\text{strain} = \frac{\delta V}{V} = \frac{\text{stress}}{B} = \frac{1.1 \times 10^8\text{ Pa}}{2.1 \times 10^9\text{ Pa}} = 0.052.$$

But $\delta V/V = \delta \rho/\rho$ (by conservation of mass) so the density at the trench bottom is

$$\rho_b = (1025\text{ kg m}^{-3}) \times (1 + 0.052) = 1080\text{ kg m}^{-3}.$$

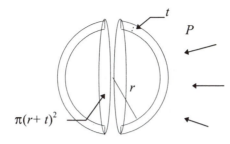

Figure 56. A spherical bathyscaph

We can therefore neglect the effect of change of density on pressure.

(c) If the sphere has inner radius r and thickness t the net force pushing the two hemispheres together horizontally is

$$\text{Force} = \text{pressure} \times \text{normal area} = P\pi(r+t)^2.$$

The area of steel supporting this force is $2\pi rt$. So the stress is

$$\text{Stress} = \frac{P\pi(r+t)^2}{2\pi rt}.$$

This produces a compressive strain in the steel. To estimate the thickness we can use Young's modulus since this is very nearly the same as the bulk modulus. For the maximum allowed strain assume a value of 0.005 so it is still in the Hookean regime. So

$$\text{Strain} = \frac{P(r+t)^2}{2rt} \times \frac{1}{Y} = 0.005,$$

from which, for $t \ll r$,

$$t \simeq \frac{Pr}{2 \times 0.005Y} = \frac{(1.1 \times 10^8 \text{ Pa}) \times (1 \text{ m})}{0.01 \times (2 \times 10^{11} \text{ Nm}^{-2})} \sim 0.055 \text{ m}.$$

So, about $\boxed{6 \text{ cm}}$ is the minimum thickness required to withstand the pressure.

Tutorial 19   Hot and cold

19.1   Physics of dieting

(a) A body at temperature T loses heat to surroundings at temperature T_0 at a net rate $\sigma(T^4 - T_0^4)$ per unit area per unit time. Take $T = 310$ K (corresponding to body temperature of 37 C) and $T_0 = 288$ K. To obtain the total surface area A we approximate the body by a cylinder of 1m circumference and height 1.75m. The area of the cylindrical surface is 1.75×1 m^2. The area of the top and base is $2\times(1/4\pi)$ m^2 which is smaller and can be neglected. Thus, the net rate of loss of heat is

$$A\sigma \times 4T_0^3 \Delta T \;=\; (1.75\text{ m}^2)\times(5.7\times10^{-8}\text{J m}^{-2}\text{K}^{-4}\text{s}^{-1})\times4\times(288\text{ K})^3\times22\text{ K}$$
$$\approx\; 210\text{ J s}^{-1}.$$

For 1 day ($\sim10^5$s), we require an intake of at least $\boxed{20\text{ MJ}}$.

(b) If a kg of fat provides about 20 MJ then not eating consumes about 1 kg of fat per day, so this is the minimum rate of mass loss.

19.2   Saving the *Titanic*?

Take the iceberg to be a slab of constant cross-section A and of mass M.

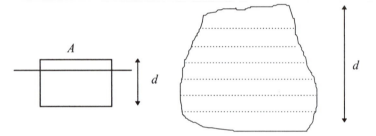

Figure 57. An irregular iceberg can be considered as a number of cylindrical layers

If the latent heat of melting of ice is L, and the solar constant is S then the time t to melt a depth d of the iceberg is given by

$$SAt = LA\rho d$$

from which the time to melt a depth of 1m is

$$t = \frac{L\rho d}{S} = \frac{(333\times10^3\text{J kg}^{-1})\times(10^3\text{ kg m}^{-3})\times(1\text{m})}{1360\text{ Wm}^{-2}} = 2.5\times10^5\text{ s m}^{-1}.$$

For an iceberg 30 m thick the minimum time is 100 days (1 day $\sim10^5$s). Note that the time to melt depends only on the number of layers to be melted. It is therefore independent of the shape and depends only on the depth of the iceberg.

The effect of the soot is not large (a factor 2 at most) but the time gained is significant in preventing the iceberg from entering the shipping lanes.

19.3   Winter sports

(a) The density of water is a maximum at 4 °C. So water at this temperature in a cold pond sinks to the bottom and stays there even when water above is cooled below 4 °C by the air.

(b) The heat of fusion of water at 0 °C flows along the temperature gradient in a layer of ice of thickness x at a rate $\kappa \Delta T/x$, where κ is the thermal conductivity of the ice. The heat that has to be removed from a layer of water of thickness δx just below the ice to freeze it is $L\rho\delta x$.

Therefore, in a time δt

$$L\rho\delta x = \frac{\kappa \Delta T}{x}\delta t$$

from which, integrating and putting $x = 0$ at $t = 0$,

$$x^2 = \frac{2\kappa t\Delta T}{L\rho}.$$

This result was first obtained by Stefan in 1891. For the problem in hand,

$$x^2 = \frac{2 \times (2 \text{ W K}^{-1}\text{m}^{-1}) \times (12 \times 3600 \text{ s}) \times (10 \text{ K})}{(3.33 \times 10^5 \text{ J kg}^{-1}) \times (1000 \text{ kg m}^{-3})}$$

$$= 5.2 \times 10^{-3} \text{ m}^2$$

or $\boxed{x = 7.2 \text{ cm}}$.

(c) Using $x^2 \propto t$ and that 7.2 cm of ice forms in 12 hours, the time T required to form 30 cm is given by

$$\left(\frac{7.2}{30}\right)^2 = \left(\frac{12 \text{ hrs}}{T \text{ hrs}}\right)$$

so

$$T = 12 \times \left(\frac{30}{7.2}\right)^2 \times \frac{1}{24} \text{ days} = \boxed{8.7 \text{ days}}.$$

The lesson of this calculation is that it takes longer than you might suppose to form a safe thickness of ice because x grows only as $t^{1/2}$.

19.4 Warming the Earth

(a) In material of thermal conductivity κ heat H flows down a thermal gradient $\Delta T/\Delta x$ through an area A at a rate

$$\frac{dH}{dt} = \kappa A \frac{\Delta T}{\Delta x} = (2 \text{ W K}^{-1}\text{m}^{-1}) \times (1\text{m}^2) \times \frac{(30 \text{ K})}{(10^3 \text{ m})} = \boxed{0.06 \text{ W m}^{-2}}$$

for the Earth's crust.

(b) The heat produced in the Earth, radius R, density ρ, assuming the radioactive content of all rock yields $\varepsilon = 4 \times 10^{-10}$ J s^{-1}kg^{-1}, is $\frac{4}{3}\pi R^3\rho\varepsilon$. To obtain the heat flux per unit area of the Earth's surface we divide by $4\pi R^2$. This gives

$$\frac{dH}{dt} = \frac{1}{3}R\rho\varepsilon = \frac{1}{3} \times (6.37 \times 10^6 \text{ m}) \times (5500 \text{ kg m}^{-3}) \times (4 \times 10^{-10} \text{ W})$$

$$= 4.67 \text{ Wm}^{-2}.$$

The actual heat flow across the Earth's surface from part (a) is considerably smaller. Given that the time for heat to flow through the Earth is less than the radioactive half-lives of the elements involved, the heat flow must be in a steady state, so losses must balance the rate of production. The conclusion therefore is that the radioactive isotopes of U, Th and K must be less abundant in the core than in the crust.

Tutorial 20   Cooling the chunnel

(a) When running at maximum capacity there are 8 trains at any instant in a tunnel, so the rate of delivery of heat to the tunnel is

$$8 \times 4 \times 10^6 \text{ W}$$

If the mass rate of flow of water is $\dot{M}$ the energy is extracted at a rate

$$\Delta T \times C \times \dot{M}$$

where $C = 4190 \text{ J kg}^{-1}\text{K}^{-1}$ is the specific heat of water. For $\Delta T = 25$ C this gives

$$\dot{M} = \frac{32 \times 10^6 \text{ W}}{(25 \text{ K}) \times (4.19 \times 10^3 \text{J kg}^{-1}\text{K}^{-1})} \approx \boxed{300 \text{ kg s}^{-1}}.$$

(b) Since the refrigerator is assumed ideal the energy supplied equals the work done in cooling the water. Thus, the working substance of the refrigerator takes in heat Q at the current water temperature T, and rejects heat $Q+W$ at the temperature T_0 of the surroundings. To complete a cycle the working substance of the ideal refrigerator goes through two adiabatic transformations in which the entropy is constant. Since the cycle is reversible there is no overall change in entropy so

$$\frac{Q}{T} = \frac{Q+W}{T_0}.$$

For a mass of water M

$$Q = -MCdT,$$

So, if the initial water temperature is T_0 and the final temperature T_f then

$$-MC \int_{T_0}^{T_f} \frac{dT}{T} = -\frac{MC}{T_0} \int_{T_0}^{T_f} dT + \int_0^{W_f} \frac{dW}{T_0}$$

or

$$-MC \log_e \frac{T_f}{T_0} = -\frac{MC}{T_0}[T_f - T_0] + \frac{W_f}{T_0},$$

and hence $E = W_f$ is

$$E = -MCT_0 \log_e \frac{T_f}{T_0} - MC(T_0 - T_f).$$

(c) Let $\Delta T = T_0 - T_f \ll T_0$. Then, substituting for ΔT

$$\begin{aligned} E &= -MCT_0 \log_e(T_0 - \Delta T)/T_0 - MC\Delta T \qquad (29) \\ &= -MCT_0[\log(1 - \Delta T/T_0)] - MC\Delta T, \end{aligned}$$

where we are now ready to use the approximation $\log(1 - x) \simeq -x - \frac{1}{2}x^2$ for small $x = \Delta T/T_0$. Thus

$$\begin{aligned} E &\simeq -MCT_0\left[-\frac{\Delta T}{T_0} - \frac{1}{2}\left(\frac{\Delta T}{T_0}\right)^2 \right] - MC\Delta T \\ &= \frac{MC}{2}\frac{(\Delta T)^2}{T_0}. \end{aligned}$$

(d) From equation (29) the required power (energy per unit time) is

$$\begin{aligned} P &= \frac{\dot{M}C}{2}\frac{(\Delta T)^2}{T_0} \\ &= \frac{(300 \text{ kg s}^{-1}) \times (4190 \text{ J kg}^{-1}\text{K}^{-1}) \times (15 \text{ K})^2}{2 \times (293 \text{ K})} \\ &= \boxed{2.6 \text{ MW.}} \end{aligned}$$

Tutorial 21   Heat flow in ice-houses

(a) The contents of the ice-house are lost through melting at a rate $\dot{M}$, say. Part of the molten water flows down the drain giving a rate of loss $\dot{M}_M$, and part evaporates to the atmosphere giving a rate of loss $\dot{M}_V$.

Conservation of mass implies

$$\dot{M} = \dot{M}_M + \dot{M}_V.$$

By conservation of energy the total rate at which energy flows into the ice is

$$\dot{E} = \dot{M}_V(L_V + L_M) + \dot{M}_M L_M,$$

where L_M and L_V are the latent heat of melting and evaporation respectively. Eliminating $\dot{M}_M$ between these equations gives the *ice-house equation*

$$\dot{M} = \frac{\dot{E} - \dot{M}_V L_V}{L_M}. \qquad (30)$$

Clearly the rate of melting is reduced by increasing the evaporation.

(b) In the first case equation (30) gives the rate of melting as

$$\dot{M}_1 = \dot{E}/L_M.$$

In the second case $\dot{M} = \dot{M}_V$, which, inserted into (30) gives

$$\dot{M}_2 = \dot{M}_V = \frac{\dot{E}}{L_V + L_M}.$$

So the ratio is

$$\frac{\dot{M_1}}{\dot{M_2}} = \frac{L_V + L_M}{L_M} = \frac{(0.33 + 2.26) \text{ J kg}^{-1}}{0.33 \text{ J kg}^{-1}} = \boxed{7.85}.$$

The message of this result is that an ice-house should not be air-tight.

The point was not always appreciated by builders of ice-houses in the past, in part of course, because the ideas of latent heat and conservation of energy were not formulated until the second half of the nineteeth century.

(c) Let a fraction f of the melted ice evaporate. Then

$$\dot{M_V} = f\dot{M}.$$

Substituting into the ice-house equation (30) and rearranging gives

$$\dot{M} = \frac{\dot{E}}{L_M + fL_V} = \frac{\dot{E}}{L}, \tag{31}$$

say. Energy flows into the ice through the walls and through the air above the top surface. The latter will be small and can be neglected. Therefore, if κ is the rate of heat conduction (in W m^{-2} K^{-1}), ΔT the temperature difference between the ground and the surface of the ice, and A the surface area of the chamber in contact with the ice, and h the depth of ice, then (31) can be written

$$\frac{dM}{dt} = -\frac{\kappa \Delta T}{L} \left(2\pi rh + \pi r^2\right).$$

Since $M = \pi r^2 h\rho$ this becomes

$$\frac{dM}{dt} = -\frac{\kappa \Delta T}{L} \left(\frac{2M}{\rho r} + \pi r^2\right) = -(a + bM)$$

where $a = (\pi r^2 \kappa \Delta T)/L$, $b = (2\kappa \Delta T)/(L\rho r)$ and $a/b = m = \pi r^3 \rho/2$.

Integrating,

$$\int_{M_0}^{M} \frac{dM}{a + bM} = -\int_0^t dt$$

or

$$\frac{1}{b} \log_e \frac{(a + bM)}{(a + bM_0)} = -t$$

giving, finally

$$M = M_0 e^{-bt} - m(1 - e^{-bt}). \tag{32}$$

Now,

$$M_0 = \pi \left(\frac{3.5}{2} \text{ m}\right)^2 \times (5 \text{ m}) \times (800 \text{ kg m}^{-3}) = 38 \text{ tonnes}$$

and

$$m = \frac{\pi r^3 \rho}{2} = \frac{\pi}{2} \times \left(\frac{3.5}{2} \text{m}\right)^3 \times (800 \text{ kg m}^{-3}) = 6.7 \text{ tonnes}.$$

Thus, the second term in (32) can be neglected for small t. (The equation is valid for times t such that $M \geqslant 0$ i.e for $t < b^{-1}\log_e(1 + M_0/m)$.)

For $t \lesssim b^{-1}$ the ice mass obeys

$$\boxed{M = M_0 e^{-bt}.}$$

(d) So its half-life is

$$t_{1/2} = \frac{\log_e 2}{b} = \frac{prL}{2\kappa\Delta T}\log_e 2.$$

If there is no evaporation ($f = 0$) then $L = L_M$ and

$$t_{1/2} = \frac{(800 \text{ kg m}^{-3}) \times (\frac{3.5}{2} \text{ m}) \times (0.33 \times 10^6 \text{ J kg}^{-1})}{2 \times (1 \text{ W m}^{-2} \text{ K}^{-1}) \times 5 \text{ K}} \times \log_e 2$$

$$= \boxed{3.2 \times 10^7} \text{ s,}$$

so the half-life is about a year. The half-life would be doubled if

$$L = L_M + fL_V = 2L_M,$$

i.e. if

$$f = \frac{L_M}{L_V} = \frac{0.33 \times 10^6}{2.26 \times 10^6} \approx \frac{1}{7}.$$

So if 1/7 of the melted ice evaporates the half-life is about 2 years.

An ice house can therefore keep half its load for about a year, so a supply of ice can be maintained from one winter to the next. A well designed ice house in which evaporation is encouraged could continue to supply ice into a second year in the event of a mild winter. This agrees with the recorded performance of ice houses.

Tutorial 22   Atmospheres

22.1   Columns of air

(a) The pressure P equals the mass per unit area of the column of air $\times g$, so the mass M above an area A is

$$M = \frac{P}{g} \times A = \frac{1.0 \times 10^5 \text{ N m}^{-2}}{9.81 \text{ m s}^{-2}} \times (2 \times 10^{-2} \text{ m}^2) = \boxed{204 \text{ kg}},$$

where we have taken the area of the head to be $\pi(8 \text{ cm})^2 \approx 2 \times 10^{-2} \text{ m}^2$.

(b) The external pressure is balanced by an equal internal pressure.

(c) The height h of a unit area of liquified nitrogen must contain the same mass as the air column. Therefore

$$h = \frac{M}{A\rho} = \frac{P}{\rho g} = \frac{1.0 \times 10^5 \text{ N m}^{-2}}{(800 \text{ kg m}^{-3}) \times (9.81 \text{ m s}^{-2})} = \boxed{12.7 \text{ m}}$$

(d) There is an upper limit to the pressure at ground level because eventually the atmosphere will start to liquify.

22.2   Climbing Everest

(a) For an isothermal atmosphere of molecules of molecular mass m in hydrostatic equilibrium

$$P = P_0 e^{-mgz/kT} = P_0 \exp\left(-\frac{z}{z_0}\right),$$

where the scale height (taking an average molecular weight for air of $\mu = 28.97$) is

$$z_0 = \frac{kT}{mg} = \frac{(1.38 \times 10^{-23} \text{ J K}^{-1}) \times (288 \text{ K})}{28.97 \times (1.66 \times 10^{-27} \text{ kg}) \times (9.81 \text{ m s}^{-2})} = 8424 \text{ m}.$$

Hence $P/P_0 = e^{8428/8848} = \boxed{0.386}$. i.e. the pressure is 38.6% of that at sea level. This suggests that survival is possible, although not with a large margin of safety.

(b) For the hydrostatic equilibrium of a layer of atmosphere, thickness dz density ρ we have

$$dP = -\rho g dz$$

with equation of state

$$\rho = \frac{mP}{k(T_0 - \alpha z)}.$$

Therefore

$$\int_{P_0}^{P} \frac{dP}{P} = -\int_0^z \frac{mg dz}{k(T_0 - \alpha z)}.$$

where the lower limit is determined by $P = P_0$ at $z = 0$. Integrating gives

$$\log_e P - \log_e P_0 = \frac{mg}{k\alpha} \log_e(T_0 - \alpha z) - \frac{mg}{k\alpha} \log_e T_0$$

and, finally,

$$P = P_0 \left(1 - \frac{\alpha z}{T_0}\right)^{mg/k\alpha}.$$

(c) With $\alpha = 6.5$ K km^{-1}, $mg/k\alpha = 5.26$ so $P/P_0 = (1 - \frac{6.5 \times 8.85}{288})^{5.26} = \boxed{0.31}$. Thus, the more exact theory gives 31% of sea level pressure. (This is confirmed by standard atmospheric tables.)

Survival at the altitude of Everest without extra oxygen appears not to be possible except briefly. Everest has been climbed without oxygen: presumeably climbers do not linger at the summit.

Tutorial 23 Atmospheric friction

(a) The lowest impact velocity would be the escape velocity of the Earth. This is

$$v_{min} = \left(\frac{2GM_E}{r_E}\right)^{1/2} = \left(\frac{2 \times (6.67 \times 10^{-11} \text{ N m}^2\text{kg}^{-2}) \times (6 \times 10^{24} \text{ kg})}{6.37 \times 10^6 \text{ m}}\right)^{1/2}$$

$$= \boxed{11.2 \text{ km s}^{-1}.}$$

The maximum impact velocity would be a head on encounter with a long period comet which has very nearly the solar system escape velocity from the Earth's orbit. This is

$$v = \left(\frac{2GM}{R}\right)^{1/2} = \left[\frac{2 \times (6.67 \times 10^{-11} \text{ N m}^2\text{kg}^{-2}) \times (2 \times 10^{30} \text{ kg})}{1.5 \times 10^{11} \text{ m}}\right]^{1/2}$$

$$= 42 \text{ km s}^{-1}.$$

So the maximum velocity, $v_{max} = (30 + 42)$ km s^{-1} = $\boxed{72 \text{ km s}^{-1}.}$

In practice speeds in the range 10.8 to 39 km s^{-1} have been observed (*Physics of Meteoritic Phenomena*, V. A. Bronshten). We shall take 20 km s^{-1} as typical.

(b) This is an example of the square-cube law.

The mass of the meteorite is proportional to volume, ie mass $\propto r^3$, for a meteorite of radius r.

The surface area through which the meteorite is heated $\propto r^2$.

So the mass available to be ablated increases faster than the surface area through which it can be heated. At some radius, the transfer of heat will be insufficient to ablate the meteorite completely.

(c) Assuming the meteorite is moving normal to the Earth's surface, the total kinetic energy of the air impinging on the meteorite is given by

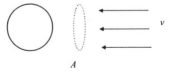

Figure 58. Impact of air flow in the frame of the meteroite

$$\text{Total K.E.} = \frac{1}{2}(\rho_0 Ah)v^2,$$

where ρ_0 is the density of the atmosphere at the ground, and h is the scale height of the atmosphere.

Let f be the heat transfer coefficient and L the heat of ablation. Then a fraction f of this kinetic energy heats the meteorite of mass M and ablates ΔM, so

$$L \Delta M = f \times \frac{1}{2}(\rho_0 A h)v^2,$$

or

$$\frac{\Delta M}{M} = \frac{f \rho_0 h A v^2}{2LM}.$$

To estimate this, we have

$$\rho_0 h = \frac{P_0}{g} = \frac{10^5 \text{ Pa}}{9.81 \text{ m s}^{-2}} = 10.2 \times 10^3 \text{ kg m}^{-2}$$

and

$$M = \frac{4}{3}\pi R^3 \rho_M = \frac{4}{3}\pi \times (6000 \text{ m})^3 \times (2500 \text{ kg m}^{-3}) = 2.3 \times 10^{15} \text{ kg}.$$

This gives

$$\frac{\Delta M}{M} = \frac{0.02 \times (10.2 \times 10^3 \text{ kg m}^{-2}) \times \pi(6000 \text{ m})^2 \times (2 \times 10^4 \text{ m s}^{-1})^2}{2 \times (5 \times 10^6 \text{ J kg}^{-1}) \times (2.3 \times 10^{15} \text{ kg})}$$

$$= \boxed{4.0 \times 10^{-4}}.$$

Thus, even for an oblique trajectory, which would increase the mass in the column of air by up to a factor of 30, the mass loss would still be small.

Note this assumes v^2 is constant. This is justified below. But in any case it cannot alter the conclusion since a decrease in v would reduce the mass loss still further.

(d) In unit time a column of air of length v, mass $\rho A v$, impinges on the body. The momentun carried by this column is $(\rho A v) \times v = \rho A v^2$. A large fraction

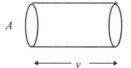

A

v

Figure 59. This column of air stikes the meteorite in a unit time

of this transferred to the body giving rise to the drag force. The drag force is proportional to the momentum transferred per unit time, hence proportional to $\rho A v^2$.

(e) The drag force on the meteorite is $\frac{1}{2}C_d \rho_{air} A v^2$, so the deceleration is

$$a = \frac{1}{2} \frac{C_d \rho_{air} A v^2}{M}.$$

Take $\rho_{air} = \rho_0$, the density at ground level, and $v = v_i$ the initial velocity to obtain an overestimate of the deceleration. For a typical value of $C_d = 1$ and

using $M = 2.3 \times 10^{15}$ kg from part (c) this gives
$$a = \frac{1 \times (1.2 \text{ kg m}^{-3}) \times \pi(6000 \text{ m})^2 \times (2 \times 10^4 \text{ m s}^{-1})^2}{2 \times (2.3 \times 10^{15} \text{ kg})}$$
$$= 12 \text{ m s}^{-2}.$$
The time spent decelerating is of order
$$t_{decel} \sim \frac{\text{scale height of the atmosphere}}{\text{speed}}$$
$$= \frac{8.5 \text{ km}}{20 \text{ km s}^{-1}} \approx 5 \text{ s}.$$
The change in speed in this time is of order $v_i t = 60 \text{ m s}^{-1} \ll v_i$. This is an overestimate (because we overestimated the deceleration) so we can assume that a 12 km asteroid would hit the ground with its initial mass and velocity. Note that the acceleration of gravity is of the same order as a so it would make little difference to the velocity and we are justified in ignoring it.

(f) The pressure P on the front face is force/area or
$$P = \frac{1}{2}C_d\rho v^2.$$
This will be a maximum at sea level (since the speed is essentially constant), so
$$P = 0.5 \times 1 \times (1.2 \text{ kg m}^{-3}) \times (2 \times 10^4 \text{ m s}^{-1})^2 = \boxed{240 \text{ MPa.}}$$
So the 12 km meteorite could well be intact when it hits the ground. If it does break it will do so in the lower atmosphere, leaving little time for the fragments to spread, so the crater size should not be affected.

(g) The kinetic energy per unit mass is $\frac{1}{2}v_i^2$ (since the impact velocity is close to v_i) or $\frac{1}{2} \times (2 \times 10^4 \text{ m s}^{-1})^2 = \boxed{2 \times 10^8 \text{ J kg}^{-1}.}$

This is a much higher figure than the yield of high explosives which is around 10^7 J kg^{-1}.

(h) The energy, E, released on impact is ~100 times the energy required to vapourise the meteorite. Shock waves will break up and heat up a volume of material at least 100 times that of the meteorite. The radius of the crater formed will therefore be $\sim (100)^{1/3} \times 6 \approx 25$ km. across. The volume of the crater formed will be proportional to E and its radius to $E^{1/3}$.

(i) The density of water is about 10^3 that of air, so, from part (f), the pressure on the meteorite as it impacts with the water will be of order 240×10^3 MPa. This is much greater than the maximum crushing strength of 500 MPa. In addition, the deceleration of the fragments would be greater by the same factor ρ_{water}/ρ_{air}, so they would not penetrate to an appreciable depth.

Tutorial 24   Bernoulli's theorem

24.1   Mir

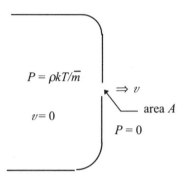

$P = \rho k T / \overline{m}$

$v = 0$

$\Rightarrow v$

area A

$P = 0$

Figure 60. The Mir problem

If gas of density ρ escapes with velocity v the mass loss rate from the hole of area A, is

$$\frac{dM}{dt} = -\rho v A$$

Bernoulli implies

$$P + 0 = 0 + \frac{1}{2}\rho v^2$$

so

$$v^2 = \frac{2P}{\rho} = \frac{2kT}{\overline{m}}$$

from the ideal gas law, $P = \rho k T / \overline{m}$, for molecules of mean molecular mass $\overline{m}$. Hence

$$\frac{dM}{dt} = -\rho A \left(\frac{2kT}{\overline{m}}\right)^{1/2},$$

where $\rho = M/V$.

Thus

$$\frac{dM}{dt} + \lambda M = 0, \tag{33}$$

where

$$\lambda = \frac{A}{V}\left(\frac{2kT}{\overline{m}}\right)^{1/2}. \tag{34}$$

We can use this relation to find A once we know λ which we get from the solution of equation (33) and the given data. The solution to the differential equation (33)

is $M = M_0 e^{-\lambda t}$ (with M_0 the initial mass of air) and hence
$$P = P_0 e^{-\lambda t}.$$
The time for the pressure to fall from 750mm of mercury to 675 mm of mercury was 8 minutes, so
$$675 = 750 e^{-\lambda \times 8 \times 60}.$$
Thus,
$$\lambda = \frac{1}{8 \times 60} \times \log_e \frac{750}{675} = 2.2 \times 10^{-4} \text{s}^{-1}.$$
We now use (34) with $T = 298$ K, $V = 390$ m^3, and $\overline{m} = 28 \times 1.66 \times 10^{-27}$kg.
$$A = \frac{V\lambda}{\left(\frac{2kT}{m}\right)^{1/2}} = \frac{(390 \text{ m}^3) \times (2.2 \times 10^{-4}\text{s}^{-1})}{\left(\frac{2 \times (1.38 \times 10^{-23} \text{ J K}^{-1}) \times (298 \text{ K})}{28 \times 1.66 \times 10^{-27} \text{ kg}}\right)^{1/2}} \text{ m}^2 = \boxed{2 \times 10^{-4} \text{ m}^2}.$$
This is about half the area of a postage stamp as reported in the press at the time.

24.2   Aerodynamic lift

(a) The wing gives the air a downward component of velocity $v_\perp$ thereby exerting a downward force on the air. By Newton's third law the reaction force on the wing is upwards.

(b) Air of density ρ is carried away from the wing at speed $v_\perp$. The downward flux of momentum is therefore proportional to $\rho v_\perp^2$. So this is the force per unit area on the wing and the lift force is $\rho v_\perp^2 A$ for a wing of area A.

(c) The upward force on the wing is $A\Delta P$ and this must balance the weight mg of the plane. Thus
$$\Delta P = \frac{mg}{A} = \frac{(3.3 \times 10^5 \text{ kg}) \times (9.81 \text{ ms}^{-2})}{(500 \text{ m}^2)} = \boxed{6.47 \times 10^3 \text{ N m}^{-2}}.$$
Atmospheric pressure is about $P_{at} = 10^5$ Nm^{-2} so $\Delta P = 0.065 P_{at}$.

(d) From Bernoulli's theorem the pressure difference across the wing is
$$\Delta P = \frac{\rho}{2}(v_u^2 - v_l^2),$$
where v_u is the speed over the upper surface and v_l that over the lower one. We express this in terms of the mean air speed.
$$\Delta P = \frac{\rho}{2}(v_u + v_l)(v_u - v_l) = \rho v \Delta v,$$
in an obvious notation. Thus
$$\frac{\Delta v}{v} = \frac{\Delta P}{\rho v^2} = \frac{(0.065 \times 10^5 \text{ kg m}^2)}{(1.2 \text{ kg m}^{-3}) \times (267 \text{ m s}^{-1})},$$
corresponding to a speed of 600 mph with the density of air taken at sea level. Thus $\boxed{\Delta v / v = 0.076}$ so the speed over the top of the wing is about 7% faster

than under the bottom.

A common explanation of flight, that the longer arc of wing on the top surface forces air to speed up which causes a drop in pressure clearly cannot be correct since planes can fly upside down. It is also rather strange that a change in velocity (i.e. an acceleration) should be the *cause* of a force! In reality, the shape and attitude of the wing redirect the momentum of the air downwards and give rise to pressure gradients above the wing that bring about an increase in the flow speed - as they must do for consistency with Bernoulli's theorem. As usual, it is not possible to make causal deductions from a conservation law such as Bernoulli's.

24.3   The flight of a golf ball

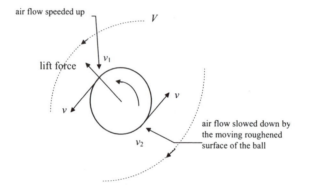

Figure 61. Air flow in the rest frame of the ball

In the absence of spin the air flow is symmetrical so there is no lift. For a spinning ball take the air speed above the ball to be $v_1 = V + v$ and the air speed below the ball to be $v_2 = V - v$ (the maximum amounts by which the flow can be affected by the spin of the ball, since not all points on the ball rotate with the equatorial velocity).

Bernoulli's theorem gives

$$P_1 + \frac{1}{2}\rho v_1^2 = P_2 + \frac{1}{2}\rho v_2^2.$$

For a ball of cross section area A, the lift force is

$$\text{Lift} = (P_2 - P_1)A = \frac{1}{2}\rho A(v_1^2 - v_2^2) = \frac{1}{2}\rho A(v_1 - v_2)(v_1 + v_2) = 2\rho AvV.$$

From the data

$$v = a\omega = (2.1 \times 10^{-2} \text{ m}) \times (30 \times 2\pi \text{ rad s}^{-1}) = 3.96 \text{ m s}^{-1}.$$

Thus,

Lift $= 2 \times (1.2$ kg m$^{-3}) \times \pi (2.1 \times 10^{-2}m)^2 \times (3.96$ m s$^{-1}) \times (50$ m s$^{-1}) = \boxed{0.66 \text{ N}}$.

The weight of the ball is $mg = (0.046$ kg$) \times (9.81$ m s$^{-2}) = 0.45$N. So the lift to weight ratio is about 1.5. This is close to the experimentally determined value (*The Search for the Perfect Golf Swing*, J Stobbs & A Cochran).

Tutorial 25   Spontaneous structure

25.1   Ordering life?

(a) The free energy per particle is given in terms of the entropy per particle s and energy per particle u as $f = u - Ts$ for a system at temperature T. The only difference between the perfect gas and the system under consideration is in the addition to the internal energy. Thus

$$f = f_{gas} + \varepsilon.$$

Alternatively, to derive this from first principles, consider the N-particle partition function for a gas with internal energy

$$Z = \frac{1}{N!} \left(\iint \frac{d^3p\, d^3x}{h^3} \exp \left(-\frac{p^2}{2mkT} + \frac{\varepsilon}{kT} \right) \right)^N.$$

This is

$$Z = Z_{gas} e^{N\varepsilon/kT}$$

where Z_{gas} is the partition function for a perfect gas. The free energy per particle is $f = (kT \log_e Z)/N = f_{gas} + \varepsilon$ as required.

(b) The number of aggregates of size N is P_N/N per molecule (surfactant + water). So their contribution to the free energy per particle $f_N \propto P_N/N$. Since the volumes occupied by a water molecule and by a surfactant molecule are assumed to be equal, the total volume $\propto$ total number of surfactant and water molecules. Thus, the number of aggregates per unit volume n having aggregation number N is also proportional to P_N/N. Finally, the aggregates of different aggregation number are distinguishable, so the total free energy is the sum of the individual contributions:

$$F = \sum_N \frac{P_N}{N} \left\{ kT \left[\log_e \left(a\frac{P_N}{N} \right) - 1 \right] + \varepsilon_N \right\}$$

where a is a constant.

(c) This is to be minimised with respect to P_N subject to a fixed number of solute molecules,

$$\phi = \sum_N P_N = \text{constant.}$$

Thus (by the method of Lagrange multipliers) we minimise $\mathcal{F} = F - \mu\phi$. This gives

$$\frac{\partial \mathcal{F}}{\partial P_N} = \frac{kT}{N} \log_e \left(a\frac{P_N}{N} \right) + \frac{kT}{N} - \frac{kT}{N} + \frac{\varepsilon_N}{N} - \mu = 0$$

from which

$$P_N = aNe^{N(\mu - E_N)}$$

where $E_N = \varepsilon_N / NkT$.

(d) Consider now the case when $N = 1$ or M with $\varepsilon_M < \varepsilon_1$ (the condition for aggregation). The multiplier μ is determined by the constraint

$$\phi = P_1 + P_M = k(e^{\mu - E_1} + Me^{M(\mu - E_M)}).$$

As $\phi \to 0$, $\mu \to -\infty$ and $P_M \ll P_1$ (since $e^{M\mu} \ll e^\mu$). As μ increases (gets less negative) ϕ increases (towards 1). As μ approaches E_M then $M(\mu - E_M)$ becomes small while $\mu - E_1$ remains relatively large (since $E_M < E_1$). Thus for ϕ near 1, $P_1 \gg P_M$. There is a critical ϕ_c where μ satisfies

$$e^{\mu - E_1} = Me^{M(\mu - E_M)}.$$

We conclude that at sufficiently high concentrations $\phi > \phi_c$ the molecules spontaneously aggregate.

25.2   Oil on troubled water

(a) For instability we require

$$\rho' U^2 > \rho \left(\frac{g\lambda}{2\pi} + \frac{2\pi T}{\rho\lambda} \right) \tag{35}$$

for some wavelengths λ. For small wavelengths and for large wavelengths the right hand side of (35) is large, so there is some wavelength at which it is a minimum. This is given by

$$\lambda = 2\pi\sqrt{\frac{T}{\rho g}}.$$

In this case we require

$$U^2 > \frac{2\rho}{\rho'} \sqrt{\frac{gT}{\rho}}. \tag{36}$$

Thus

$$U^2 \gtrsim \frac{2}{1.26 \times 10^{-3}} \times \left(\frac{(9.81 \text{ m s}^{-2}) \times (7.4 \times 10^{-2} \text{ N m}^{-1})}{1.02 \times 10^3 \text{ kg m}^{-3}} \right)^{1/2} = 42.3 \text{ m s}^{-1}.$$

or $\boxed{U = 6.51 \text{ m s}^{-1}}$.

(b) The condition (36) implies that ripples will occur at lower wind speed as T is reduced. If ripples occur on water they would therefore be expected to occur under the same conditions on oil.

(c) Since the surface tension of the underlying water is constant, ripples on the oil-water interface would cause relative movement between the oil film and the water which involves the breaking of molecular bonds. This produces a friction at the surface which damps the incipient ripples sufficiently to prevent their growth. Note that the oil film cannot damp the long wavelength, deep water waves.

Tutorial 26   Material strength

26.1   All cracked up

(a) The breaking stress is the force per unit area required to dissociate the metallic bonds. This is of the order of the interaction energy per unit volume. The energy per atom is $(3.0 \text{ eV}) \times (1.6 \times 10^{-19} \text{ J eV}^{-1}) = 4.8 \times 10^{-19}$ J per molecule. The unit cell contains 4 atoms and has edge $\sqrt{2} \times 2.88 \times 10^{-10}$ m$= 4.07 \times 10^{-10}$ m. Therefore the volume per atom is

$$\text{volume/atom} = \frac{1}{4} \times (4.07 \times 10^{-10})^3 \text{m}^3 = 1.69 \times 10^{-29} \text{m}^3.$$

The energy per unit volume is therefore

$$(4.8 \times 10^{-19} \text{ J/molecule})/(1.69 \times 10^{-29} \text{ m}^3/\text{molecule}) = 2.84 \times 10^{10} \text{ N m}^{-2}.$$

This is the predicted breaking stress for a perfect crystal. The breaking stress of silver wire is $\boxed{2.9 \times 10^8 \text{ N m}^{-2}}$, a factor of about 100 less.

(b) The strain energy per unit area of the plate is $\frac{1}{2} \times$ stress $\times$ strain $\times$ unit thickness $\sim \frac{1}{2}\sigma_0 \times (\sigma_0/Y)$. The crack relieves this stress over an area having length l and some orthogonal distance from the crack. Since there is no other relevant length scale in the problem this distance must itself be of order l. Thus the decrease is strain energy is of order $l^2\sigma_0^2/2E$.

The detailed solution based on the calculated stress distribution on the plate given by Griffith produces an additional factor of $\pi/2$.

(c) The increase in the crack by a length δl produces two surfaces of area equal to (plate thickness) $\times \delta l$, and hence requires an energy $2T\delta l$ per unit plate thickness. In order that the crack can propagate this must be provided by the change in strain

energy, so
$$\delta \left(\frac{l^2 \sigma_0^2}{2Y} \right) = 2T\delta l.$$

(d) Integrating gives
$$l = \frac{4YT}{\sigma_0^2} = \frac{4 \times (7 \times 10^{10} \ \text{Nm}^{-2}) \times (0.56 \ \text{Nm}^{-1})}{(7 \times 10^7 \ \text{Nm}^{-2})^2} = \boxed{3.2 \times 10^{-5} \ \text{m}}$$
or about 0.03 mm.

26.2   Building in brick

The maximum height occurs when the weight of the building per unit area exceeds the compressive strength. Thus
$$h = \frac{\text{compressive strength}}{\rho g} = \frac{(1 \times 10^6 \ \text{Nm}^{-2})}{(2000 \ \text{kg m}^{-3}) \times (9.81 \ \text{ms}^{-2})} = \boxed{51 \ \text{m}}$$
is the height of the tallest brick building.

The tallest brick building appears to be the Moradnock built in the 1880s in Chicago which had 16 stories and walls 1.5 m thick at the lower floors. Taller building can be constructed in stone (for which the theoretical limit is the same as the maximum height of a mountain or about 5 miles, the height of Everest) or using steel girders. The first steel frame building was constructed in the US in the 1880s.

26.3   Up in smoke

The energy per unit mass required to vapourise a plane comprises the heat required to raise the temperature to the boiling point plus the heat of vapourisation. The second of these is relatively small, so we consider only the first. For an aluminium plane (boiling point 2400K) this is
$$C_v \Delta T = (880 \ \text{J kg}^{-1} \ \text{K}) \times 2400 \ \text{K} = 2.1 \times 10^6 \ \text{J kg}^{-1}.$$
The kinetic energy per kilogram is $\frac{1}{2}v^2$ so the plane must have a speed in excess of
$$v = (2 \times 2.1 \times 10^6 \ \text{J kg}^{-1})^{1/2} = 2050 \ \text{m s}^{-1},$$
which would be about Mach 4. This is not possible.

26.4   Flowing glass

Near the top of the pane the problem is the converse of the usual flow in a pipe in that for the window the surface is free but the central plane is constrained to have zero velocity. Let the z-axis be oriented upwards in the central plane of the glass and let x measure the distance from the central plane. For a layer of glass

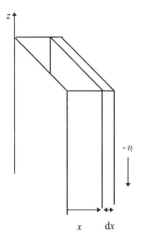

Figure 62. Flowing glass

of thickness δx at a distance x from the central plane the shear force is $\rho g dx$ and the difference in the viscous forces acting at the boundaries is $\delta(\eta dv_z/dx)$. The equation of motion is therefore

$$\eta \frac{d^2 v_z}{dx^2} = -\rho g$$

with the boundary condition $v_z = 0$ at $x = 0$ and $\frac{dv_z}{dx} = 0$ at $x = a$, the semi-thickness of the pane. Integrating gives

$$\eta \frac{dv_z}{dx} = (a - x)\rho g$$

which expresses the stress on the plane at a distance x from the centre line due to the weight of glass between the plane and the face of the window. The solution (obtained by considering $x > 0$ and $x < 0$ separately) is

$$v_z = -\frac{\rho g}{2\eta}(x^2 - 2a|x|).$$

Note that v_z is symmetrical about $x = 0$ but $dv_z/dx \neq 0$ at $x = 0$. (In fact, it is not defined at this point.)

The mass flux per unit length is

$$2 \int_0^a \rho v_z dx = -\frac{2}{3}\rho^2 g \frac{a^3}{\eta}$$

so one quarter, say, of the mass in a height $h = 10$ cm for a window of thickness 0.5 cm will flow in a time

$$t = \frac{0.25 \times \rho a h}{\frac{2}{3}\rho^2 g \frac{a^3}{\eta}} = \frac{3\eta h}{8\rho a^2 g}$$

$$= \frac{3 \times \eta \times (0.1 \text{ m})}{8 \times (1000 \text{ kg m}^{-3}) \times (0.25 \times 10^{-2} \text{ m})^2 \times (10 \text{ m s}^{-2})}$$

$$= 0.6\eta \text{ s} \sim 2 \times 10^{-8}\eta \text{ yrs.}$$

Thus we require $\eta \sim 2.5 \times 10^{10}$ N s m^{-2} if the glass is to flow in less than 500 years. Hence

$$\log_e 2.5 \times 10^{10} = 23.94 = -9.94 + \frac{17962}{T}$$

or

$$T = \frac{17962}{23.94 + 9.94} = \boxed{530 \,°\text{C}}.$$

Note that for this temperature the calculation is self-consistent. Unless one can detect much smaller flows the appearance of old windows is more likely to be a reflection of how they were made.

Tutorial 27   Electric shocks

27.1   Electric sight

(a) The n resistors in parallel have a resistance $R_T = R/n$ so

$$I = \frac{V}{\left(\frac{R}{n} + r + r_i\right)}.$$

The power in the light bulbs $= I^2 R_T$.

The power in each bulb is

$$\frac{I^2 R_T}{n} = \frac{I^2 R}{n^2} = \frac{V^2}{\left(\frac{R}{n} + r + r_i\right)^2 n^2} = \frac{R}{R} \frac{V^2}{R} \frac{1}{\left(1 + \frac{n(r+r_i)}{R}\right)^2}.$$

If $R/n \ll r + r_i$ then Preece's conclusion holds: the power of each bulb decreases like $1/n^2$.

(b) Power $= V^2/R$ so, for a modern 60 W light bulb, $R = (240 \text{ V})^2/60 \text{ W} = \boxed{960 \,\Omega}$. (In fact, the resistance of a light bulb depends strongly on temperature and is lower when cold, hence the tendancy for light bulbs to fail just when they are switched on.)

(c) For n not too large, $R/n \gg r + r_i$, so Preece's conclusion is not valid. In fact, under these conditions, to a good approximation the power per bulb is independent of n. The number of bulbs required to reduce the power per bulb from V^2/R to $\frac{1}{4}V^2/R$ is given by $n(r + r_i) = R$, or $\boxed{n \sim 1000}$ if we take $r + r_i \sim 1/5 \,\Omega$.

(d) (i) Extra generators can be switched on. (ii) High transmission voltages

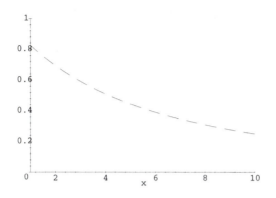

Figure 63. $y = $ *(power dissipation in a lamp)/*(V^2/R) *as a function of* $x = $
(number of lamps) $\times (r + r_i)/10R$

(400 kV) are used. A given power VI can then be provided with a lower current which reduces the transmission losses I^2R in the line.

27.2   Working at high potential

(a) The potential of the line is changing sinusoidally so until they are connected there will be a potential difference between the linesman and the line. The field is higher at an equipotential surface of higher curvature, so electrical breakdown of the air occurs preferentially at points with highest radius of curvature.

(b) For a sphere with charge Q at a potential V the capacity is $C = Q/V$. The potential at the sphere equals that of a point charge Q at its centre. So for a sphere of radius a, $V = Q/(4\pi\varepsilon_0 a)$ and $C = 4\pi\varepsilon_0 a$.

To estimate a, an average 70 kg linesman has a volume $\frac{4}{3}\pi a^3 = (70/1000)$ m^3 from which $a = 0.26$ m.

Hence $C = 4\pi\varepsilon_0 a = 4\pi \times (8.85 \times 10^{-12}$ F m$^{-1}) \times (0.26$ m$) = \boxed{29 \text{ pF}}$.

(c) The capacitance C of the linesman with respect to the rest of the universe is his capacity in isolation plus that with respect to the Earth and other bodies. However, these are relatively small so can be neglected. The linesman's potential changes at the line frequency w so his impedence is $Z = R + 1/(iwC)$. Since R the resistance of the line connecting him to the power line is small

$$|Z| = \frac{1}{|iwC|} = \frac{1}{2\pi \times (50 \text{ s}^{-1}) \times (2.9 \times 10^{-11} \text{ F})} = 1.1 \times 10^8 \text{ }\Omega$$

The current is therefore $V/|Z| \approx 4 \times 10^4$ V/10^8 $\Omega = \boxed{4 \text{ mA}}$.

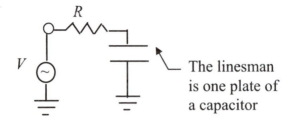

Figure 64. The linesman as capacitor

(d) This current flows through the metal suit and does no harm. The metal suit acts as a Faraday cage. Its potential goes up and down but inside it there is no electric field.

27.3 Electric defibrillation

The time constant for the circuit is $\tau = CR_{\text{total}} = C(R + r)$, where r is the internal resistance and R the trans-chest resistance. We require $r \ll R$ (in order that the power is not dissipated in the internal resistance). So
$$C = \tau/R = 20 \text{ ms}/150\Omega = \boxed{1.3 \times 10^{-4} \text{ F.}}$$
The energy in the pulse gives the required voltage: $\frac{1}{2}CV^2 = 360$ J implies $V = \boxed{2353 \text{ V.}}$

Tutorial 28   Magnetic fields

28.1   Lightning strikes

We can obtain the magnetic field in the shell from Ampere's law if we are given that the current is uniform. If the inner radius of the tube is b then at radius r
$$\int \mathbf{B}.\mathbf{dl} = \int \mathbf{j}.\mathbf{dS}$$
or
$$2\pi r B = \mu_0 I \frac{\pi(r^2 - b^2)}{\pi(a^2 - b^2)}.$$
The field is zero on the inner surface and $\mu_0 I/2\pi a$ on the outer surface. Hence
$$\boxed{B \sim \mu_0 I/4\pi a.}$$

(b) The easiest way to estimate the force is by dimensional analysis. The force between current carrying elements dl_1 and dl_2 separated by a distance r_{12} is

$\mu_0 I_1 I_2 dl_1 dl_2 / r_{12}^2$, so $\mu_0 I^2$ has the dimensions of a force. Pressure is force per unit area. The only relevant area in the problem is the cross section of the shell. If $b \ll a$ this is πa^2 so the pressure (force per unit area) is of order $F/A = \mu_0 I^2 / a^2$.

Alternatively, the force per unit volume at a point is $j \wedge B$, so the force on a sector of unit height and cross section dA is $j \wedge B \, dA$. Integrating between the inner and outer shells this gives a force per unit area (pressure) of $F/A = \mu_0 I^2 / 8\pi^2 a^2$. This can also be confirmed by direct integration of the force on one element due to another.

(c) To crush a thin copper tube requires something like a bodyweight applied to 1 cm^2 i.e 50g kg wt cm^{-2}. Therefore, if a lightening strike crushes such a tube it must produce a comparable pressure. If $a = 1$ cm this gives

$$I = \left(\frac{8\pi^2 a^2 F}{\mu_0 A} \right)^{1/2}$$

$$= (4\pi \times 10^{-7} \text{H m}^{-1})^{-1/2} \times (8\pi^2 \times 10^{-4} \text{m}^2)^{1/2} \times (5 \times 10^5 \times 9.81 \text{ kg wt m}^{-2})^{1/2}$$

$$= \boxed{1.8 \times 10^5 \text{ A}}.$$

The quoted value is around 30 000 A.

28.2 Magnetic pliers

(a) The simplest way to get at the force is from the energy. The energy of a dipole of moment $\mathbf{m}$ in a field $\mathbf{H}$ is $-\mathbf{m}.\mathbf{H}$, so the force is $\nabla(-\mathbf{m}.\mathbf{H}) = -\mathbf{m}.\nabla\mathbf{H}$.

(b) The magnetisation is $\mathbf{M} = \mu\mathbf{H}$. We have $\mathbf{H} = \mathbf{B}/\mu_0$ and, from part (a), $B = \mu_0 I / 4\pi r$. So the magnetisation (per unit volume) has magnitude

$$M = \mu B/\mu_0 = \mu I/4\pi r = \frac{10^3 \times (1.8 \times 10^5 \text{ A})}{4 \times \pi \times r} = \frac{1.4 \times 10^7}{r} \text{A m}^{-1}$$

with r in metres. The saturation magnetisation is of order 2 Bohr magnetons per atom. A Bohr magneton is $\mu_B = e\hbar/2m_e = 9.3 \times 10^{-24}$ J T^{-1}. We have $10^3 N_0/56$ atoms per kg, or $10^3 N_0 \rho / 56$ atoms m^{-3}, where the density of iron is $\rho = 7900$ kg m^{-3}. So the saturation magnetisation is

$$M = \frac{2\mu_B \times 10^3 N_0 \rho}{56}$$

$$= \frac{2 \times (9.3 \times 10^{-24} \text{J T}^{-1}) \times (6.0 \times 10^{26} \text{ Mole}^{-1}) \times (7900 \text{ kg m}^{-3})}{56}$$

$$= 5.5 \times 10^5 \text{ A m}^{-1}.$$

Thus, at distances $r < 25$ m from the lightening bolt the magnetisation saturates.

(c) For a nail length l and radius a the saturation magnetic moment is

$$m = M\pi a^2 l$$

so the force is of order

$$
\begin{aligned}
\frac{mB}{r} &= \frac{m\mu_0 I}{4\pi r^2} = \frac{2\mu_B \times 10^3 N_0\rho}{56} \times \frac{\pi a^2 l \mu_0 I}{2\pi r^2} \\
&= \frac{10^3 N_0 \mu_B \mu_0 m_{\text{nail}} I}{56\pi r^2} \\
&= \frac{1}{(56 \text{ kg Mole}^{-1})\pi} \times (6.0 \times 10^{26} \text{ Mole}^{-1}) \times (9.3 \times 10^{-24} \text{ J T}^{-1}) \times \\
&\quad \times (4\pi \times 10^{-7} \text{ H m}^{-1}) \times (1.8 \times 10^5 \text{ A}) \frac{m_{\text{nail}}}{r^2} \\
&= \frac{2.6 \, m_{\text{nail}}}{r^2} \text{ N.}
\end{aligned}
$$

with the mass of the nail, m_{nail}, in kg. If we take $m_{\text{nail}} = 10$ g then at $r = 1$ m the force is 2.6×10^{-2} N. If this force were to act in the appropriate direction, it would clearly not remove a nail, even one loosened by the torque due to the field, since the translational force is less than $m_{\text{nail}} g$. The effect is more likely to be due to currents flowing elsewhere that cause the timbers to contract.

Tutorial 29   Model circuits

29.1   The CR circuit

(a) To move a charge dQ across a voltage V_b requires an energy (work done) $V_b dQ$. So the total energy supplied by the battery in moving a charge Q is

$$
\mathcal{E} = \int_0^Q V_b dQ = V_b Q
$$

where $Q = CV_b$ is the final charge on the capacitor in the steady state. Thus

$$
\mathcal{E} = CV_b^2.
$$

But the energy stored in a capacitor is well known to be $\frac{1}{2}CV_b^2$. We expect that the missing energy has been dissipated in the resistor. To confirm this we calculate

$$
\mathcal{E}_R = \int I^2 R dt
$$

where I is the current in the circuit at time t. For the CR series circuit we have

$$
R\dot{Q} + Q/C = V_b
$$

and, either by finding an integrating factor, or using a trial solution,

$$
Q = CV_b(1 - e^{-t/RC}).
$$

Therefore, the current builds up in time according to

$$
I = \dot{Q} = \frac{V_b}{R}e^{-t/RC}.
$$

We can now evaluate the energy dissipated as

$$\mathcal{E}_R = \frac{V_b^2}{R} \int_0^\infty e^{-2t/RC} dt = \frac{1}{2} C V_b^2$$

as anticipated.

(b) As $R \rightarrow 0$ the charging timescale RC tends to zero. i.e. the charge is transported instantaneously across the system. This is obviously incompatible with relativity. In fact, (i) the current distribution becomes non-uniform round the circuit and the assumption of steady currents breaks down; and (ii) the rapid acceleration of charge in the system leads to radiation losses which act like a resistance. In practice, therefore, steady current theory breaks down and there is effectively an irreducible non-zero resistance. The full Maxwell equations would be needed to analyse the problem.

29.2   The LCR circuit

(a) The equation of motion for a series LCR circuit is

$$L\ddot{Q} + R\dot{Q} + Q/C = V_b, \tag{37a}$$

an equation which appears in many other contexts in physics. The general solution is

$$Q = CV_b + Ae^{\lambda_1 t} + Be^{\lambda_2 t} \tag{38}$$

with

$$\lambda_1 = -\frac{R}{2L} - \sqrt{\frac{R^2}{4L^2} - \frac{1}{CL}} ; \qquad \lambda_2 = -\frac{R}{2L} + \sqrt{\frac{R^2}{4L^2} - \frac{1}{CL}}.$$

The current is given by

$$I = \dot{Q} = \lambda_1 Ae^{\lambda_1 t} + \lambda_2 Be^{\lambda_2 t}. \tag{39}$$

We obtain A and B by requiring $Q = 0$ and $I = 0$ at $t = 0$.

Obviously if we fix $I = 0$ at $t = 0$ for any L, it will be zero for the particular value $L = 0$. The easiest way to show that we can fix $I = 0$ (i.e. that this initial condition is compatible with the differential equation) is to exhibit the explicit solution. This will also show us what goes wrong at $t = 0$. So, imposing $Q = 0$ and $I = 0$ at $t = 0$, we find

$$A + B = -CV_b,$$
$$\lambda_1 A + \lambda_2 B = 0,$$

from (38) and (39) respectively. The solution is $A = \lambda_2 CV_b/(\lambda_1 - \lambda_2)$ and $B = -\lambda_1 CV_b/(\lambda_1 - \lambda_2)$.

(b) For small L then

$$\lambda_1 = -\frac{R}{2L} - \frac{R}{2L}\sqrt{1 - \frac{4L}{R^2 C}} \rightarrow -\frac{R}{L},$$

$$\lambda_2 = -\frac{R}{2L} + \frac{R}{2L}\sqrt{1 - \frac{4L}{R^2 C}} \rightarrow -\frac{1}{RC},$$

so the current is

$$I \simeq \frac{\lambda_1 \lambda_2 C V_b}{\lambda_1 - \lambda_2} e^{\lambda_1 t} - \frac{\lambda_1 \lambda_2 C V_b}{\lambda_1 - \lambda_2} e^{\lambda_2 t}$$

For small, but finite L, as $t \rightarrow 0$ this gives

$$I \rightarrow \frac{\lambda_1 \lambda_2 C V_b}{\lambda_1 - \lambda_2}((1 + \lambda_1 t) - (1 - \lambda_2 t)) = \frac{V_b}{L}t.$$

So $I = 0$ at $t = 0$ for any small finite value of L.

(c) For $L = 0$ it is clear this solution is not valid (for example $\lambda_1 \rightarrow \infty$). The correct result is obtained by solving equation (37a) with $L = 0$. We have

$$R\dot{Q} + Q/C = V_b,$$

and, either by finding an integrating factor or using a trial solution,

$$Q = CV_b(1 - e^{-t/RC}),$$

with

$$I = \dot{Q} = \frac{V_b}{R}e^{-t/RC}.$$

The current now has the non-zero value V_b/R at $t = 0$.

The strange result comes about because setting $L = 0$ changes the character of the differential equation governing the system - from second order to first order - and hence the character of the solutions. (Mathematically, there are two limits involved here, $t \rightarrow 0$ and $L \rightarrow 0$ and the order in which these are taken *does* matter.) In a practical circuit the current cannot achieve a finite value in zero time. Either the real circuit will have a small but non-zero inductance which cannot be neglected for a short time after the circuit is made, or, if the inductance really is negligible, then for a short initial period steady current theory will not apply.

The moral of this story is not that models cannot be trusted, but that a physicist needs to know the assumptions that go into a formula or a result before it can be used with confidence. Usually this means knowing how the result has been derived.

Tutorial 30   Dispersion on the line

30.1   Thomson's speechless cable

(a) The solution corresponds to a wave with exponentially decreasing amplitude, frequency ω and wavenumber $k = (\omega RC/2)^{1/2}$. The phase velocity is therefore

$$v = \frac{\omega}{k} = \frac{\omega}{\left(\frac{\omega RC}{2}\right)^{1/2}} = \left(\frac{2\omega}{RC}\right)^{1/2}.$$

The velocity is a function of ω so the waves are dispersive.

(b) The object is to show that dispersion will smear out speech by causing relative propagation delays that exceed a period of oscillation. For a 1 kHz signal the time to cross the Atlantic is

$$\Delta t_1 = \frac{3600 \times 10^3 \text{m}}{\left(\frac{2 \times 2\pi \times 10^3 \text{ Hz}}{(7 \times 10^{-3} \ \Omega\text{m}^{-1}) \times (7.5 \times 10^{-11} \text{ F m}^{-1})}\right)^{1/2}} = 2.3 \times 10^{-2} \text{ s.}$$

For a 2 kHz signal the delay is $\Delta t_2 = \Delta t_1/\sqrt{2} = 1.6 \times 10^{-2}$ s. The difference in the delays is 7×10^{-3} s. The periods of 1 and 2 kHz waves are $\tau_1 = 10^{-3}$ s and $\tau_2 = 0.5 \times 10^{-3}$ s, much less than the relative delay. Thus, dispersion will smear out speech.

(c) At a frequency of 2 Hz the transit time across the Atlantic is 0.5 s, so at this frequency the cable would effectively be carrying one pulse at a time. With sufficient sensitivity it should be possible to detect the pulses. Morse code operates at a comparable frequency.

(d) Differentiating with respect to k the dispersion relation $k = (\omega RC/2)^{1/2}$, from part (a), we get

$$1 = \frac{1}{2}\left(\frac{RC}{2\omega}\right)^{1/2}\frac{d\omega}{dk}.$$

The group velocity is therefore

$$v_g = \frac{d\omega}{dk} = \left(\frac{8\omega}{RC}\right)^{1/2},$$

and this is the speed of transport of signal energy. We find $v_g > c$ for

$$\omega > \frac{c^2 RC}{8} = \frac{(3 \times 10^8)^2 \times (5 \times 10^{-13})}{8} = 5.6 \times 10^3 \text{ rad s}^{-1},$$

or a frequency $f \gtrsim 1000$ Hz.

30.2   Heaviside - speaking

(a) For $L\omega/R \ll 1$ we obtain the lowest order approximation to γ^2 by neglecting

Lw/R compared to unity:

$$\gamma^2 = \left(\frac{RC\omega}{2}\right)\left[\left(1 + \frac{L^2\omega^2}{R^2}\right)^{1/2} - \frac{L\omega}{R}\right]$$

$$\simeq \frac{RC\omega}{2}.$$

Then

$$k = \frac{RC\omega}{2\gamma} \simeq \frac{RC\omega}{2\left(\frac{RC\omega}{2}\right)^{1/2}} = \left(\frac{RC\omega}{2}\right)^{1/2},$$

and we are back to Thomson's solution.

(b) We have $L\omega/R \ll 1$ if

$$f = \frac{\omega}{2\pi} \ll \frac{R}{2\pi L} = \frac{7 \times 10^{-3}\ \Omega\text{m}^{-1}}{2\pi \times 4.6 \times 10^{-7}\ \text{Hm}^{-1}} = 2.4 \times 10^3\ \text{Hz},$$

which is adequate for the transmission of Morse code.

(c) Now consider the high frequency case, $L\omega/R \gg 1$. Expanding by the binomial theorem we have

$$\gamma^2 = \frac{1}{2}RC\omega\left[\frac{L\omega}{R}\left(1 + \frac{R^2}{L^2\omega^2}\right)^{1/2} - \frac{L\omega}{R}\right]$$

$$\simeq \frac{1}{2}RC\omega\left[\frac{L\omega}{R} + \frac{R}{2L\omega} - \frac{L\omega}{R}\right]$$

$$\simeq \frac{R^2C}{4L},$$

which is independent of ω, and $v = \omega/k = 2\gamma/RC \simeq (LC)^{-1/2}$, which is also independent of ω. So dispersion is no longer a problem.

The natural inductance of the cables was not high enough at the frequencies that were used, so to raise $\omega L/R$ it was necessary to increase L by inserting inductors at intervals along the line. Heaviside also showed that if the line has a conductance G across the dielectric then the condition $RC = LG$ produces exactly zero dispersion.

(d) We start from

$$v = \frac{\omega}{k} = \frac{2\gamma}{RC},$$

which shows that the speed v is a maximum at the maximum value of γ. This occurs for $\omega \to \infty$, in which case $\gamma \to (R^2C/4L)^{1/2}$. Thus $v_{\text{max}} = (LC)^{-1/2}$.

Now, using the values for a coaxial line, $LC = \mu\mu_0\varepsilon\varepsilon_0 = \mu\varepsilon/c^2$. Thus $v = c/(\mu\varepsilon)^{1/2}$ and attains its maximum value of c, the speed of light, for a vacuum between the inner and outer conductors.

Tutorial 31   Things to do with mirrors

31.1   Solar furnaces

(a) Assume that we can approximate the parabolic mirror as the arc of a circle. Then the linear size of the image is αD (figure 65).

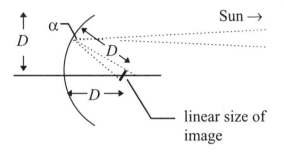

Figure 65. The image formed at the focus of an approximately circular mirror

(b) The solar constant at the Earth is
$$S = \frac{L}{4\pi R^2} \text{ where } L = 4\pi a^2 \sigma T_s^4.$$
So $S = a^2 \sigma T_s^4 / R^2$. But (figure 66) $\alpha = 2a/R$, so
$$S = \frac{1}{4}\alpha^2 \sigma T_s^4.$$

(c) The energy per second through the image
$$= \quad \text{the energy collected from the Sun}$$
$$= \quad \text{solar constant} \times \text{area of mirror}$$
$$= \quad \frac{1}{4}\alpha^2 \sigma T_s^4 \pi D^2.$$

Since we treat the image as a spherical ball the incident light falls normally on the surface and we do not have to worry about projected areas (as we would for the image on a flat screen). So the energy incident per second on unit area averaged over the spherical surface of the image is
$$\frac{\frac{1}{4}\alpha^2 \sigma T_s^4 \pi D^2}{4\pi(\frac{\alpha D}{2})^2} = \frac{\sigma T_s^4}{4}.$$
This expression is independent of distance from the Sun.

(d) To collect all the light from the Sun place a black body at the focus of such a size as just to fill the image. In equilibrium this body, radius r, receives as much

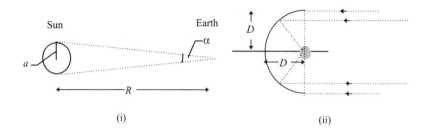

Sun   Earth

a α

R

(i)   (ii)

Figure 66. (i) The angular size of the Sun from the Earth and (ii) The image is illuminated over half its surface by rays incident normally

energy as it radiates, so, if it is at a temperature T_{bb},

$$4\pi r^2 \frac{\sigma T_s^4}{4} = 4\pi r^2 \sigma T_{bb}^4. \tag{40}$$

Thus, $T_{bb} = T_s/4^{1/4} = 0.71 T_s$.

(e) There would be a conflict with the second law of thermodynamics if the blackbody were hotter than the Sun. In this case heat would flow from a cooler to a hotter body without work being done. The left hand side of (40) is correct for any (axially symmetric) shaped object (by Gauss's theorem and conservation of flux) and the minimum value for the energy radiated (the right hand side) is $\pi r^2 \sigma T_{bb}^4$ for a disc. In this case the temperature of the disc would equal that of the Sun.

(f) At some distance from the Sun the diffraction limit is reached and the mirror can no longer resolve the Sun's disc. For light of wavelength λ this occurs when $\alpha = 1.22\lambda/D$. At this point the image size becomes independent of distance and the argument in (b) no longer applies.

For $\alpha = 2a/R = 1.22\lambda/D$ we have

$$R = \frac{2aD}{1.22\lambda} = \frac{2 \times (6.96 \times 10^8 \text{m}) \times (3\text{m})}{1.22 \times (500 \times 10^{-9}\text{m})} = \boxed{6.85 \times 10^{15}\text{m}}.$$

which is less than the distance to the nearest star.

31.2   Going to war with mirrors

(a) If the area of a mirror is A, and S is the solar constant, the power received at the mirror is AS. The angular diameter of the Sun is $\alpha = \frac{1°}{2} \times \frac{2\pi}{360} = 8.7 \times 10^{-3}$ radians. This is focussed by a mirror of focal length $f = 50$ m on to an image of radius $\alpha f/2$. The power per unit area at the vessel is therefore

$$\frac{AS}{\pi \left(\frac{\alpha f}{2}\right)^2}.$$

This is required to exceed 6.3×10^4 W m^{-2}. The solar constant is 1360 W m^{-2}. The actual incident power of the Sun even in summer in Greece is less than this. For an order of magnitude take $S = 1000$ W m^{-2}. Putting in the values, we need

$$A \geqslant \frac{(6.3 \times 10^4 \text{ Wm}^{-2})}{1000 \text{ Wm}^{-2}} \times \pi \left(\frac{8.7 \times 10^{-3} \text{ rad} \times 50 \text{ m}}{2}\right)^2 = 9.4 \text{ m}^2$$

or a mirror of radius $\boxed{1.7 \text{ m}}$.

(b) With a parabolic mirror the solar flux is concentrated by a factor (area of mirror/area of image) $\gg 1$. With plane mirrors this factor is < 1 and the light is diluted. For simplicity consider a circular plane mirror. If the mirror diameter is d the image at a distance f of a source subtending an angle α is

$$\frac{\pi d^2}{4} + \pi d \times \frac{\alpha}{2} \times f = \frac{\pi \times (1.128 \text{ m})^2}{4} + \pi(1.128 \text{ m}) \times \frac{8.7 \times 10^{-3}}{2} \times 50 \text{ m}$$

$$= 1.77 \text{ m}^2$$

for a mirror of area 1 m^2 at a distance of 50 m. So the dilution factor is $1\text{m}^2/1.77$ m$^2 \approx 1/2$. Thus each plane mirror contributes at most 1/2 of a solar constant. Therefore $6.3 \times 10^4/(1000/2) = 126$ mirrors are required.

A more detailed calculation gives of order 440 mirrors of area 1 m^2 (Mills A. A., *E.J.Phys.*, 1992, **13**, 268). The story is therefore probably false.

Tutorial 32 Blackbody radiation

32.1 'Blackish bodies'

(a) To be an exact blackbody a system must be in thermal equilibrium. There can therefore by no flow of heat (no temperature gradients). On the other hand in order to radiate a net energy flux to its environment a body must possess a thermal gradient. An isolated body cannot therefore radiate exactly as a blackbody. Approximate blackbodies are obtained for systems with negligible net fluxes. This can be either because the system has been artificially constructed to trap radiation (as in Kirchoff's box) or because radiation is trapped in the body by absorption and scattering processes. (In this case the system is said to be optically thick.) So radiation is trapped for a long time in the interior of the Sun, the thermal gradients are small and the condition for an approximate blackbody are satisfied. With the exception of some absorption lines that are formed at lower temperatures in the overlying chromosphere, the optical emission comes to us from the photosphere, a thin layer of approximately constant temperature. (The photosphere is the effective visible surface of the Sun.)

In contrast a gas flame is optically thin so radiation can escape freely down a large temperature gradient. The flux is large compared to the energy density and the flame does not approximate a blackbody.

There is also a more fundamental explanation in terms of microscopic physics. In thermal equilibrium there must be a detailed balance (equilibrium) between the rates at which energy is being transferred between subsytems. In the gas flame this is not the case: the atoms are excited (gain energy from other atoms, or free electrons) by collisions but are de-excited (lose their energy to photons) by radiative transitions.

(b) The flux radiated by a source subtending a solid angle $d\Omega$ in a frequency band ν to $\nu+d\nu$ is given by the Rayleigh-Jeans approximation to the blackbody spectrum:

$$F_\nu d\nu = \frac{2\nu^2}{c^2} kT d\nu d\Omega.$$

To obtain the flux in a band we should integrate over frequency. This gives an overall frequency dependence $\int_{\nu_0-\Delta\nu}^{\nu_0+\Delta\nu} \nu^2 d\nu \simeq \nu_0^2 \Delta\nu$. For the purpose of an order of magnitude estimate we use the fact that a radio receiver operates with a bandwidth $\Delta\nu$ of order 8 kHz . So for a blackbody at $T = 1000$ K at radio frequencies, $\nu_0 \approx 2 \times 10^5$ Hz, the flux is

$$\frac{2}{(3 \times 10^8 \text{ m s}^{-1})^2} \times (2 \times 10^5 \text{ Hz})^2 \times (8000 \text{ Hz}) \times (1.4 \times 10^{-23} \text{ J K}^{-1}) \times$$
$$\times (1000 \text{ K}) \times \Delta\Omega$$
$$\sim 10^{-22}\Delta\Omega \text{ W m}^{-2}.$$

A gas flame, which is not a blackbody must radiate less than this. Therefore any gas flame radiates less than about 10^{-22} W m^{-2} at kHz radio frequencies. This is not detectable.

(c) A thermal source of angular scale 10^{-3} seconds of arc radiating a spectral flux $F_\nu = 10^{-29}$ W m^{-2}Hz^{-1} would have a temperature

$$T = \frac{F_\nu c^2}{2\nu^2 k \Delta\Omega} = \frac{10^{-29} \times (3 \times 10^8)^2}{2 \times (10^9)^2 \times (1.4 \times 10^{-23}) \times (10^{-3} \times \frac{2\pi}{360 \times 3600})^2} \sim 10^9 \text{ K}.$$

This is an unrealistically high temperature for a radio source. (The emission from a thermal plasma at this temperature would be dominated by $\gamma-$radiation.)

The temperature calculated in this way is called the *brightness temperature* of the source.

32.2   Photon numbers

(a) The simplest way to get the number density of blackbody photons is to divide the energy density, $u = aT^4$, by the average photon energy $h\bar{\nu} = 2.7kT$. We have

$$n_{room} = \frac{aT^4}{2.7kT} = \frac{(7.56 \times 10^{-16}\ Jm^{-3}K^{-4}) \times (293\ K)^3}{2.7 \times (1.38 \times 10^{-23}\ J\ K^{-1})}$$

$$= \boxed{5.1 \times 10^{14}\ \text{photons m}^{-3}}.$$

For the air, $P = nkT$ gives

$$n_{air} = \frac{(10^5\ N\ m^{-2})}{(1.38 \times 10^{-23}\ J\ K^{-1}) \times 293\ K} = \boxed{2.47 \times 10^{25}\ \text{molecules m}^{-3}}.$$

(b) For the universe

$$n_{Universe} = n_{room} \times \left(\frac{T_{Universe}}{T_{room}}\right)^3$$

$$= (5.1 \times 10^{14}\ m^{-3}) \times \left(\frac{2.7}{293}\right)^3$$

$$= 3.4 \times 10^8\ \text{photons m}^{-3}$$

and, since the Universe is mainly hydrogen

$$n_{matter} = \frac{\rho}{m_p} = \frac{10^{-26}\ kg\ m^{-3}}{1.67 \times 10^{-27}\ kg} \approx 6\ \text{protons m}^{-3}.$$

So universally photons dominate over material particles.

(c) Inside LEP the photon density will be the same as in (a) because the temperatures are the same. The number of air molecules will scale with the pressure. The number is therefore reduced by a factor $10^{-10}/760 = 1.3 \times 10^{-13}$, so the number density of molecules is $3.25 \times 10^{12}\ m^{-3}$ which is less than the photon density.

(d) The electrons and positrons scatter off of the photons giving up energy in the process to produce γ-rays. (This is sometimes called inverse Compton scattering.)

32.3   The Sun as a blackbody

(a) Yellow light has a wavelength $\sim$500 nm so from Wien's law corresponds to the blackbody peak for a temperature,

$$T = \frac{2.9 \times 10^{-3}\ K\ m}{500 \times 10^{-9}\ m} = 5800\ K.$$

We measure the energy flux from the Sun at the Earth (the solar constant) to be

$s = 1360$ W m^{-2}. If the Earth-Sun distance is D then the total power of the Sun is $L = 4\pi D^2 s$. From Stefan's law, the luminosity from a blackbody of area $4\pi R^2$ at temperature T is $L = 4\pi\sigma R^2 T^4$. Therefore

$$R = \left(\frac{L}{4\pi\sigma}\right)^{1/2}\frac{1}{T^2} = \left(\frac{4\pi \times (1360\ \text{W m}^{-2}) \times (1.5 \times 10^{11}\ \text{m})^2}{4\pi \times (5.67 \times 10^{-8}\ \text{W m}^{-2}\text{K}^{-4})}\right)^{1/2}\left(\frac{1}{5800}\right)^2$$

$$= \boxed{6.91 \times 10^8\ \text{m}}.$$

The true radius is 6.96×10^8 m. Historically this calculation was first used in 1897 to find the solar temperature once Stefan's law had been proposed. Other suggested laws of cooling had given results in the range 1300 K to 10^6 K.

(b) The method applies equally to any object which can be approximated as a blackbody radiator.

For Betelgeuse the result is around 3×10^{11} m which means that if it replaced the Sun it would encompass the orbit of Mars. This result (proposed around the turn of the century) was not finally accepted until the radius of Betelgeuse was measured in 1921.

32.4   The Jupiter star

(a) At closest approach Jupiter is 4.5 au from the Earth (where 1 au = the mean Earth-Sun distance = 1.5×10^8 km).

The luminosity of the Jupiter star (J_*) would be

$$L_J = 4\pi R_J^2 \sigma T^4 = 4\pi \times (7 \times 10^7\ \text{m})^2 \times (5.67 \times 10^{-8}\ \text{W m}^{-2}\text{K}^{-4}) \times (5800)^4$$
$$\approx 4 \times 10^{24}\ \text{W}.$$

The flux at the Earth would be

$$F_J = \frac{L_J}{4\pi R_J^2} = \frac{4 \times 10^{24}\ \text{W}}{4\pi \times (4.5\ \text{au} \times 1.5 \times 10^{11}\ \text{m au}^{-1})^2} = \boxed{0.7\ \text{W m}^{-2}}.$$

(b) This can be compared with the flux F_S from the Sun which is 1360 Wm^{-2}. The flux of energy F_J from J_* would be negligible, but it would be extremely bright. If Ω_J and Ω_S are the solid angles subtended by J_* and the Sun respectively, the ratio of brightnesses (energy fluxes per unit solid angle) would be

$$\frac{F_J/\Omega_J}{F_S/\Omega_S} = \frac{L_J/R_J^2}{L_S/R_S^2} = \frac{4 \times 10^{24}\ \text{W}}{3.9 \times 10^{30}\ \text{W}} \times \left(\frac{7.0 \times 10^8\ \text{m}}{7.0 \times 10^7\ \text{m}}\right)^2 \approx 10^{-4}.$$

This is greater than the surface brightness of the Moon. The apparent astronomical magnitude of J_* would be $m_J = -18.62$ compared with $m_{\text{Moon}} = -12.7$ for full Moon, and its energy flux $F_J = F_{\text{Moon}} \times 10^{0.4(m_{\text{Moon}} - m_J)} \approx 100$. Thus the night sky would not be dark when the Jupiter star was above the horizon.

Tutorial 33   Relativistic travel

(a) There is no limit *in principle,* i.e. no kinematic limit, to how far a person can travel in a lifetime. Travelling at a speed v close to c, the distance travelled in a lifetime, τ is

$$D = \gamma v \tau \qquad (41)$$

The velocity cannot exceed c but γ can be made as large as we please, so a finite τ is no bar to travel anywhere.

(b) The time dilation formula relates galaxy frame time t to spaceship time τ,

$$t = \gamma \tau. \qquad (42)$$

Doing the calculation exactly, from (41) we have

$$\gamma v = D/\tau.$$

and, as $\gamma = (1 - v^2/c^2)^{-1/2}$, $v = c(\gamma^2 - 1)^{1/2}/\gamma$ so

$$\gamma = \left\{ \left(\frac{D}{c\tau}\right)^2 + 1 \right\}^{1/2} \qquad (43)$$

$$= \left\{ \left(\frac{2 \times 10^{22}}{3 \times 10^8 \times 50 \times 3.2 \times 10^7}\right)^2 + 1 \right\}^{1/2} = \boxed{42000}. \qquad (44)$$

For the case of an 8 TeV proton in the LHC we have

$$\gamma = \frac{E}{m_p c^2} = \frac{8 \times 10^{12} \text{ eV}}{930 \times 10^6 \text{ eV}} = 8600.$$

This illustrates the enormity of the γ-factor of the spaceship.

(c) Using (42), $t = \gamma \tau = \{(D/c)^2 + \tau^2\}^{1/2} = \frac{D}{c}\{1 + c^2\tau^2/D^2\}^{1/2}$. But $c\tau/D \ll 1$, so

$$t = \frac{D}{c} + \frac{1}{2}\frac{c\tau^2}{D} + \ldots$$
$$= 2109705 \text{ years} + 5.2 \text{ hours}$$

The first term is the expected light travel time and the second term represents the additional time taken by the travellers moving at less than the speed of light. The light signal arrives back on Earth after a lapse of 22,109,705 years in $\boxed{4,221,910 \text{ AD}}$. Such a project might be thought to lack the appeal of immediacy.

(d) From part (b) $\gamma = 42000$, so the distance in the spaceship frame is $D/\gamma = (2 \times 10^{22} \text{ m})/42000 = 4.76 \times 10^{17} \text{ m} \boxed{\sim 15 \text{ pc}}$.

Note that the arrival of the astronauts at their destination is attributed to time dilation from the point of view of the Earth frame but to length contraction from the viewpoint of the spaceship frame. This is a general feature of relativity: observers in different frames will always agree on what happens if they both view the same events, but they will disagree, in general, on time and length measurements and whether events are simultaneous.

Tutorial 34   Relativistic hyperbolic motion

34.1   Constant acceleration in relativity

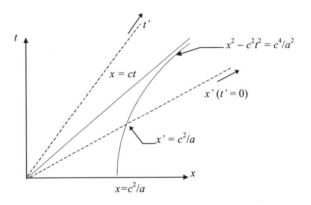

Figure 67. World line of constant acceleration

(a) The asymptote is the world line of a photon emitted from the event at the origin.

(b)The axes of the frame S' are show as dashed in the figure. The hyperbola $x^2 - c^2t^2 = c^4/a^2$ is the locus of events at an interval separation from the origin of c^2/a. At $t = 0$ in S the proper distance of the spaceship from O is c^2/a. Since the interval is invariant, the proper distance of the spaceship from the origin in any instantaneously co-moving frame S' at $t' = 0$ is $x' = c^2/a$. Therefore this distance is constant.

34.2   Siamese rockets

(a) The rockets start together in the initial rest frame and accelerate identically so the distance apart remains the same in this frame. But a moving rod of fixed length suffers length contraction, so the rest length of the cable must increase with speed to maintain constant length in S. Hence it breaks.

(b) It breaks at the start of the journey.

34.3   Rigid bodies

(a) For motion with constant acceleration we have
$$x^2 - c^2t^2 = c^4/a^2. \tag{45}$$
At $t = 0$, $x_1 = c^2/a_1$ and $x_2 = c^2/a_2$ so $x_1 - x_2 = L_0 = c^2(1/a_1 - 1/a_2)$. As $x_2 < x_1$ it follows that $a_2 > a_1$.

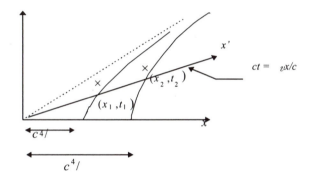

Figure 68. World lines for rigid body motion with uniform acceleration

(b) We have to (i) find the frame in which both rockets are at rest at any time and (ii) show that the spatial separation of the rockets remains constant in this frame.

We obtain the velocity of each rocket by differentiating the equation for hyperbolic motion (45) :
$$2\frac{dx}{dt}x - 2c^2t = 0$$
from which
$$v = \frac{c^2t}{x}. \tag{46}$$
Thus the line $v =$ const. is $x = (c^2/v)t$. The frame S' in which this is the x' axis (i.e. the line $t' = 0$) is the simultaneous rest frame for the two rockets.

Since $x^2 - c^2t^2 = c^4/a^2$ is a Lorentz invariant in the rest frame S' we have also $x'^2 - c^2t'^2 = c^4/a^2$. So the spatial distance between the two events at (x_1, t_1) and (x_2, t_2) in the simultaneous rest frame $t' = 0$ is $x'_1 - x'_2 = c^2/a_1 - c^2/a_2$.

This is the same as the spatial distance between the two rockets in their initial frame S, so the relative motion is rigid.

(c) Eliminating t between equations (45) and (46) we obtain the velocity as a function of position. As $x \rightarrow 0$ then from (45) the acceleration of the rear rocket $\rightarrow \infty$ and $v \rightarrow c$. (A photon has infinite proper acceleration.)

Tutorial 35   Mechanics near c

35.1   The size of particle accelerators

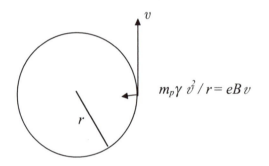

Figure 69. Charged particle orbit in a uniform magnetic field into the page

(a) The equation of motion of a proton moving at velocity v in an orbit of radius r, to which it is confined by a magnetic field B is

$$\frac{d\mathbf{p}}{dt} = eBv\hat{\mathbf{r}}$$

where $\mathbf{p} = m_p \gamma \mathbf{v}$ is the momentum of the proton.

At maximum energy the speed of the proton is constant. Also, the acceleration $\left|\frac{d\mathbf{v}}{dt}\right| = \frac{v^2}{r}$ for a circular orbit. Therefore

$$\frac{m_p \gamma v}{r} = eB$$

or $p = eBr$. For highly relativistic motion $E \gg mc^2$, and $E \simeq cp$ is, to a good approximation, the kinetic energy (i.e. the energy in excess of the rest mass energy). Thus

$$r = \frac{E}{eBc} = \frac{(8 \times 10^{12} \text{ eV}) \times (1.6 \times 10^{-19} \text{ J eV}^{-1})}{(1.6 \times 10^{-19} \text{ C}) \times (9 \text{ T}) \times (3 \times 10^8 \text{ m s}^{-1})} = \boxed{2.96 \times 10^3 \text{ m}}$$

and the circumference is 19 km.

This is an underestimate of the size of the ring because this contains straight sections to accommodate detectors etc. The LHC will be built in the LEP tunnel which is about 27 km in circumference.

(b) To complete the picture we do the Newtonian calculation. Here $p = m_p v$, so the equation of motion is

$$\frac{m_p v^2}{r} = eBv$$

which gives $p = Ber$ again. However, the kinetic energy is now $E = p^2/2m_p$ so

$$
\begin{aligned}
r &= \frac{(2m_p E)^{1/2}}{Be} \\
&= \frac{[2 \times (1.67 \times 10^{-27} \text{ kg}) \times (8 \times 10^{12} \text{ eV}) \times (1.6 \times 10^{-19} \text{ J eV}^{-1})]^{1/2}}{(9 \text{ T}) \times (1.6 \times 10^{-19} \text{ C})} \\
&= \boxed{45.4 \text{ m}} \, !
\end{aligned}
$$

Thus relativity is responsible for the high cost of building particle accelerators.

35.2   Relativistic snooker

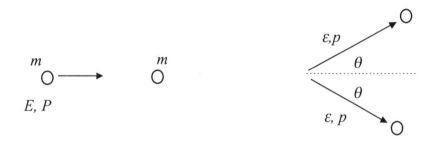

Figure 70. A relativistic collision

By symmetry, the two balls go off with equal energy and momentum after the collision. The energies and momenta are related by

$$E^2 = c^2 P^2 + m^2 c^4$$

and

$$\varepsilon^2 = c^2 p^2 + m^2 c^4.$$

Conservation of energy gives

$$E + mc^2 = 2\varepsilon. \qquad (47)$$

Conservation of momentum gives:

$$P = 2p \cos \theta.$$

So

$$\cos^2\theta = \frac{c^2 P^2}{c^2(2p)^2} = \frac{E^2 - m^2c^4}{4(\varepsilon^2 - m^2c^4)}.$$

Using (47) we get

$$
\begin{aligned}
\cos^2\theta &= \frac{E^2 - m^2c^4}{4(\frac{1}{4}(E + mc^2)^2 - m^2c^4)} \\
&= \frac{E^2 - m^2c^4}{(E^2 + 2Emc^2 - 3m^2c^4)} \\
&= \frac{(E - mc^2)(E + mc^2)}{(E - mc^2)(E + 3mc^2)}.
\end{aligned}
$$

Now compute $\tan\theta$ from

$$\tan^2\theta = \frac{1 - \cos^2\theta}{\cos^2\theta} = \frac{2mc^2}{E + mc^2} = \frac{2}{1 + E/mc^2}$$

and finally

$$\tan\theta = \left(\frac{2}{1 + E/mc^2}\right)^{1/2}.$$

For a small initial kinetic energy we recover the classical result $\theta = 45°$.

For large initial energy, $E \gg mc^2$, we have $\theta < 45°$, so the angle between the balls will be less than a right angle.

35.3   A relative paradox

The body does not slow down in S because the changing momentum comes from its changing mass, not from its changing velocity. (In Newtonian physics there is no link between the loss of internal energy and a change in mass.) First we need to find the momentum carried away by the radiation in the S frame, given that it carries zero net momentum in the S' frame. We use the Lorentz transformation for energy and momentum. In the S' frame $p'_{\text{rad}} = 0$ and the energy is E'; in the S frame the momentum is

$$p_{\text{rad}} = \gamma\left(p'_{\text{rad}} + \frac{E'v}{c^2}\right) = \gamma\frac{E'v}{c^2}.$$

Now, assume the particle has a speed u in S after the emission. Conservation of linear momentum in S requires

$$
\begin{aligned}
\gamma m v &= \gamma(m - E'/c^2)u + p_{\text{rad}} \\
&= \gamma m u
\end{aligned}
$$

from which $u = v$. Momentum is conserved and the speed of the particle is, in fact, unchanged in both frames.

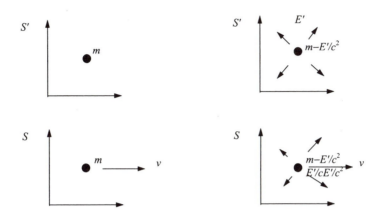

Figure 71. Emission of radiation viewed from frames in relative motion

Tutorial 36   A relativistic aberration

(a) The velocity transformations are

$$u'_x = \frac{u_x - v}{\left(1 - \frac{vu_x}{c^2}\right)}, \tag{48}$$

$$u'_y = \frac{u_y}{\gamma\left(1 - \frac{vu_x}{c^2}\right)} \tag{49}$$

where

$$u_x = c\cos\theta$$
$$u_y = -c\sin\theta,$$

and, similarly

$$u'_x = c\cos\theta'$$
$$u'_y = -c\sin\theta'.$$

Therefore, using (49)

$$-c\sin\theta' = \frac{-c\sin\theta}{\gamma(1 + \frac{v}{c}\cos\theta)},$$

or

$$\sin\theta' = \frac{\sin\theta}{\gamma(1 + \frac{v}{c}\cos\theta)} \tag{50}$$

giving the angle θ' in the S' frame.

(b) We are give $\theta = 150°$ and $\theta' = 30°$ so $\sin\theta = \sin\theta' = \sin 30° = 1/2$, and

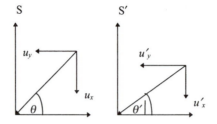

Figure 72. The frames S and S' have relative velocity v in the x-direction

$\cos 150° = -\sqrt{3}/2$. So from (50) we get

$$1/2 = \frac{1/2}{\gamma\left(1 - \frac{v}{c}\frac{\sqrt{3}}{2}\right)}$$

from which, putting $\gamma = (1 - v^2/c^2)^{-1/2}$, we get $\boxed{v = \frac{4\sqrt{3}}{7}c}$ for the required speed of the spaceship.

(c) The sin transformation (50) does not allow us to distinguish $\theta' > 90°$ from $\theta' < 90°$ so we turn to the transformation for the x-component of velocity (48). Thus

$$\cos\theta' = \frac{\cos\theta + \frac{v}{c}}{1 + \frac{v}{c}\cos\theta}. \tag{51}$$

If $\cos\theta < 0$ and $\cos\theta' > 0$ we must have $v > c|\cos\theta|$ (or, equivalently, $\sin\theta = (1 - \cos^2\theta)^{1/2} > 1/\gamma$), where $c|\cos\theta|$ is the component of velocity of the ray in the direction of motion of the spaceship. Thus the spaceship moves faster than the ray in this direction.

(d) We show that no ray with $\cos\theta < 1$ can be aberrated into the straight ahead direction. To do this we use (51). If $\theta' = 0°$ (for a ray straight ahead) then

$$1 = \frac{\cos\theta + \frac{v}{c}}{1 + \frac{v}{c}\cos\theta}$$

which has $\cos\theta = 1$ (i.e $\theta = 0°$) as the only solution.

(e) The stars near the edge of the Galaxy are aberrated to an inner ring of the image. Stars nearer the centre of the Galaxy appear further out in the spaceship frame since they are aberrated to larger angles. The scale of the image is determined by the speed of the spaceship: the larger the speed the more concentrated is the image towards the inner ring.

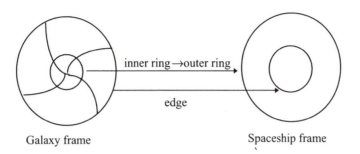

inner ring→outer ring

edge

Galaxy frame   Spaceship frame

Figure 73. The Galaxy is turned inside out by aberration

(f) At constant speed the inner edge of the image will expand outwards so the annular image of the Galaxy will become a thinner ring.

(g) Consider the motion of the spaceship under constant acceleration. The spaceship undergoes hyperbolic motion, given by

$$(x + c^2/a)^2 - c^2 t^2 = c^4/a^2, \qquad (52)$$

where $x = 0$ at $t = 0$.

We want an expression for the distance travelled as a function of γ. To obtain this we differentiate to get

$$2(x + c^2/a)u_x - 2c^2 t = 0,$$

where $u_x = dx/dt$. Re-arranging and eliminating t from the equation of the trajectory (52) we obtain

$$x = \frac{c^2}{a}(\gamma - 1).$$

At a distance x from the Galaxy, radius R, we have $\tan \theta = R/x$. Imposing the condition $\sin \theta > 1/\gamma$ from part (c) then gives

$$\frac{R}{(R^2 + x^2)^{1/2}} > \frac{1}{\gamma},$$

or

$$\frac{R}{\left(R^2 + \frac{c^4}{a^2}(\gamma - 1)^2\right)^{1/2}} > \frac{1}{\gamma}$$

from which we can derive

$$a > \frac{c^2(\gamma - 1)^{1/2}}{R(\gamma + 1)^{1/2}}$$

provided $\gamma > 1$. So, if $a > c^2/R$ (say), this condition will be satisfied at all times. For a large enough acceleration the galaxy will be aberrated into the forward direction! (The slightly odd consequence that the required aberration occurs from the start, even for small speeds, arises from the assumption that the spacecraft takes off from the plane of the Galaxy, so that initially only an infinitesimal amount of aberration is needed.)

Tutorial 37   Gravitational Lensing

(a) Figure 74 shows the relative positions of the star, observer and lens and the asymptotes to the photon trajectories at the star and the observer. Let the radius of the Einstein ring be r. The angle of deflection of a ray passing at distance r

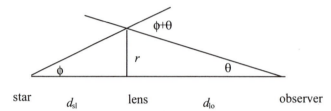

Figure 74. Geometry of lensing

from a lens of mass M is

$$\theta + \phi = \frac{4GM}{c^2 r}.$$

For small angles $\phi = r/d_{sl}$ and $\theta = r/d_{lo}$ so

$$\frac{r}{d_{sl}} + \frac{r}{d_{lo}} = \frac{4GM}{c^2 r}.$$

Let $\mu = (1 + d_{lo}/d_{sl})$; solving for r gives

$$r = \left(\frac{4GMd_{lo}}{c^2 \mu}\right)^{1/2}; \qquad \text{so } \theta = \frac{r}{d_{lo}} = \left(\frac{4GM}{c^2 \mu d_{lo}}\right)^{1/2}.$$

For a solar mass lens with $d_{lo} = 10$ kpc and $d_{sl} = 50$ kpc,

$$\theta = \left(\frac{4 \times (6.67 \times 10^{-11} \text{ N m}^2\text{kg}^{-2}) \times (2 \times 10^{30}\text{kg})}{(3 \times 10^8 \text{ ms}^{-1})^2 \times (1 + 0.2) \times (10^4 \text{ pc} \times 3.1 \times 10^{16} \text{ m pc}^{-1})}\right)^{1/2}$$

$$= \boxed{4 \times 10^{-9} \text{ radians}},$$

and $r = \theta d_{lo} \approx 10^9$km.

(b) The resolution of the 3m Hubble telescope mirror is $1.22\lambda/D = 1.22 \times (5 \times 10^{-7} \text{ m})/3 \text{ m} = 2 \times 10^{-7}$ radians for light of wavelength 500nm in the optical. So the Hubble Space Telescope cannot resolve the Einstein ring in this case. It can resolve the ring of a distant quasar lensed by a galaxy.

(c) Light rays are null geodesics in space-time. These are independent of frequency.

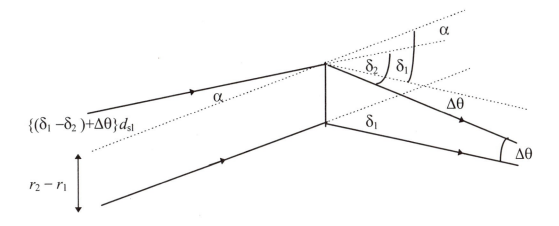

Figure 75. The lensing of an extended object. The angle α is given by $\alpha + \delta_2 = \delta_1 + \Delta\theta$.

(d) The angle subtended by the ring at the observer is
$$\Delta\theta = \theta_2 - \theta_1 = (r_2 - r_1)/d_{lo} = \Delta r/d_{lo},$$
where r_1 is the inner radius of the ring and r_2 its outer radius. From figure 75, the diameter of the star is $(r_2 - r_1) + (\delta_1 - \delta_2)d_{sl} + \Delta\theta d_{sl}$ and hence the angle subtended by the star at the observer, $\Delta\lambda$ (not shown in the figure), is
$$\Delta\lambda = \frac{\Delta r}{d} + \Delta\delta\frac{d_{sl}}{d} + \Delta\theta\frac{d_{sl}}{d}, \tag{53}$$
where $\Delta\delta = (\delta_1 - \delta_2)$. The objective is to show $\Delta\lambda = 2\Delta\theta$.

In the notation of part (a) we have
$$\delta = \frac{4GM}{c^2 r}$$
$$= \theta + \phi = \frac{r}{d_{sl}} + \frac{r}{d_{lo}},$$
and
$$\Delta\delta = \Delta(4GM/c^2 r) = -\frac{\delta}{r}\Delta r = -\left(\frac{1}{d_{lo}} + \frac{1}{d_{sl}}\right)\Delta r.$$
Putting this in (53), and using $\Delta r = d_{lo}\Delta\theta$, we find
$$\Delta\lambda = \frac{\Delta r}{d} + \left(\frac{1}{d_{lo}} + \frac{1}{d_{sl}}\right)\Delta r\frac{d_{sl}}{d} + \Delta\theta\frac{d_{sl}}{d}$$
$$= \frac{d_{lo}\Delta\theta}{d}\left(1 + \left(\frac{1}{d_{sl}} + \frac{1}{d_{lo}}\right)d_{sl}\right) + \Delta\theta\frac{d_{sl}}{d}$$
$$= 2\Delta\theta,$$
as was to be proved.

This assumes the thickness of the ring is small compared with its radius. Since $\Delta\lambda = 2R_{\text{star}}/d \approx 10^9 \text{m}/(60 \times 3.1 \times 10^{19}\text{m}) = 5.4 \times 10^{-13}$ radians $\ll \theta$ ($= 2 \times 10^{-7}$ radians from part (a)), and we have just shown that $\Delta\lambda = 2\Delta\theta$, we have $\Delta\theta \ll \theta$.

(e) Consider two sections of areas dA_1 and dA_2 of a small bundle of rays, both taken normal to the rays. At dA_2 the solid angle subtended by dA_1 is $d\Omega_2 = dA_1/r^2$ and, similarly, $d\Omega_1 = dA_2/r^2$, where r is the distance between the sections. If the intensity (or 'brightness') of the radiation (energy per unit projected area per unit time per unit solid angle) is I, then conservation of energy along the bundle requires that $IdAd\Omega = $ constant. Thus $I_1 dA_1 dA_2/r^2 = I_2 dA_2 dA_1/r^2$, from which $I_1 = I_2 = I = $ constant. Thus we arrive at the standard result that intensity is constant along a ray.

(f) The flux of light from a source (energy per unit projected area per unit time) = intensity × solid angle. Since the brightness is the same for the star and the ring, the flux ratio or amplification A, is determined by the ratio of solid angles subtended. We have, using $\sin\theta \simeq \theta$,

$$A = \frac{2\pi\theta(\Delta\lambda/2)}{\pi(\Delta\lambda/2)^2} = \frac{4\theta}{\Delta\lambda} = \frac{4\left(\frac{4GM}{c^2\mu d_{\text{lo}}}\right)^{1/2}}{2\left(\frac{R_{\text{star}}}{d}\right)}.$$

Take $M = 1M_0$ then $2GM/c^2 = 3 \times 10^3\text{m}$. Also, $d = 50$ kpc and $d_{\text{lo}} = 10$ kpc so $\mu \approx 1$. Thus, the amplification factor is

$$A \approx \frac{4 \times \left(\frac{4\times(6.67\times10^{-11}\,\text{N}\,\text{m}^2\text{kg}^{-2})\times(2\times10^{30}\text{kg})}{(3\times10^8\,\text{m}\,\text{s}^{-1})^2\times(10^4\times3\times10^{16}\text{m})}\right)^{1/2}}{2 \times \frac{10^9\text{m}}{(5\times3\times10^{20}\text{m})}} = \boxed{1.33 \times 10^4.}$$

Tutorial 38   The Universe

38.1   Olber's paradox

(a) A star, radius r_*, at a distance r subtends a solid angle

$$\delta\Omega = \frac{\pi r_*^2}{r^2}.$$

Let n be the number density of stars and, to maximise sky coverage, assume there is no overlap of stars. Then stars in a shell, distance r, thickness dr, subtend a total solid angle at the observer,

$$\Delta\Omega = \frac{\pi r_*^2}{r^2} \times n \times 4\pi r^2 dr.$$

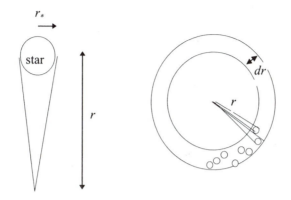

Figure 76. Stars in a shell subtend a solid angle at the observer

So the total solid angle subtended by all the stars out to $r_M = 10^{10}$ light years is

$$\Omega = 4\pi^2 r_*^2 n r_M \tag{54}$$

$$= 4\pi^2 (7 \times 10^8 \text{m})^2 \times \frac{(2 \times 10^{-26} \text{ kg m}^{-3})}{(2 \times 10^{30} \text{ kg})} \times (3 \times 10^8 \text{ m s}^{-1}) \times$$

$$\times (10^{10} \text{ lt yrs}) \times (3.15 \times 10^7 \text{ s yr}^{-1}) \tag{55}$$

$$= 1.83 \times 10^{-11} \text{ ster}$$

So the fraction of sky covered out to 10^{10} light years is $\Omega/4\pi = \boxed{1.45 \times 10^{-12}}$.

(b) For full sky coverage, still assuming no overlap, $\Omega/4\pi = 1$, we must go out to a distance (from equation (54)) of at least

$$R = \frac{r_M}{1.45 \times 10^{-12}} = \frac{3 \times 3.15 \times 10^{25} \text{ m}}{1.45 \times 10^{-12}} = \boxed{6.5 \times 10^{37} \text{ m}}.$$

The age of the universe would need to be at least the time taken for light from stars at a distance R to reach us, i.e.

$$t = R/c = \frac{(6.5 \times 10^{37} \text{ m})}{(3 \times 10^8 \text{ m s}^{-1}) \times (3.15 \times 10^7 \text{ s yr}^{-1})}$$

$$= \boxed{6.9 \times 10^{21} \text{ years}}.$$

Unless there were an extraordinary cosmic conspiracy (with stars switching on at just the right time to fill our sky with light) this is also the minimum age of stars. So the observation that stellar ages are less than 10^{11} years resolves the paradox.

(c) Let the time to fill the largest box containing just one star with radiation be t. The rate at which energy is supplied to the box in the form of blackbody radiation at temperature T from the star of area A is $A\sigma T^4$. The energy density in the filled box (at the same temperature as the star) is aT^4. So if the box has volume V

$$A\sigma T^4 t = aT^4 V$$

or

$$t = Va/A\sigma. \tag{56}$$

The volume is obtained from $V\rho = M_\odot$, where ρ is the mean mass density of the Universe and $M_\odot$ the mass of a typical star. Then, from the data in part (a)

$$V = \frac{2 \times 10^{30} \text{ kg}}{2 \times 10^{-26} \text{ kg m}^{-3}} = 10^{56} \text{ m}^{-3}.$$

So

$$t = \frac{(10^{56} \text{m}^3) \times (7.56 \times 10^{-16} \text{J m}^{-3}\text{K}^{-4})}{4\pi (7 \times 10^8 \text{m})^2 \times (5.67 \times 10^{-8} \text{W m}^{-2}\text{K}^{-4})} = 2.2 \times 10^{29} \text{s} = \boxed{6.9 \times 10^{21} \text{years}}.$$

It is, of course, no accident that this thermodynamic argument gives the same result as the geometrical one in part (b). Using $\sigma = ac/4$ in (56) gives the same algebraic expression for the time

$$t = \frac{M_\odot}{\rho} \frac{1}{\pi r_*^2 c}.$$

in both methods.

(d) Clumping of stars would lead to a partial shielding of some galaxies behind other galaxies, thereby reducing somewhat the probability of full sky coverage. This would increase the radius R to which the distribution of stars must extend to cover the sky, and hence the time required for the radiation to reach us. Clumping therefore contributes to the resolution of the paradox, but only marginally.

(e) Expansion of the Universe redshifts the light from stars so also contributes to the resolution of the paradox.

38.2   Faster than light?

(a) Let the distance d between any pair of currants be proportional to some function of time $f(t)$, the same for all pairs to ensure the expansion is homogeneous. Then

$$d \propto f(t)$$

and

$$\dot{d} \propto \dot{f}(t),$$

so

$$v = \dot{d} = \frac{\dot{f}}{f} d,$$

from which $v \propto d$.

(b) We are given

$$v = Hd = H \times \frac{2c}{H} \left[1 - (1+z)^{-1/2}\right].$$ (57)

Thus, $v \geq c$ when $2 \left[1 - (1+z)^{-1/2}\right] \geq 1$, or $\boxed{z \geq 3}$.

(c) The remotest object would have $z = \infty$, for which (57) gives $\boxed{v = 2c}$. Objects receeding at $2c$ are on our *particle horizon*.

(d) In an expanding universe particles locally at rest (the currents in the loaf) are in relative acceleration. Thus there is no global frame in which inertial observers are unaccelerated.

(e) Special relativity applies in an inertial frame. It cannot therefore be applied to an extended region of an expanding universe *even though it is valid locally*. It is easiest to understand this from the point of view of a space time diagram

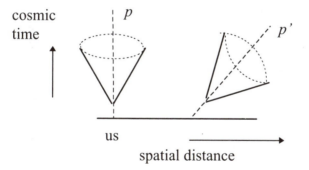

Figure 77. *Space-time diagram of two distant observers p and p' who each travel locally within their lightcones but have a relative speed $> c$*

(figure 77). The paths of particles at any event are always inside their local light cone (hence locally move at less than c). But the light cones are tilted relative to each other (by the effect of gravity on the motion of light) so that particles moving within their light cones can be moving relative to each other in excess of c.

In the Milne model of the Universe, there is expansion and a global inertial frame. However, in this system there are no particle horizons and all relative motions take place at less than c (and there is no gravity).

38.3   Gamma-ray bursts

The volume of space out to $z = 2$ at the present time is

$$\frac{4}{3}\pi d^3 = \frac{4\pi}{3}\left[\frac{2c}{H}\left(1 - \frac{1}{\sqrt{3}}\right)\right]^3.$$

So the number of galaxies out to $z = 2$ for a given number density of galaxies n (assuming that galaxies are neither created nor destroyed) is

$$
\begin{aligned}
N_g &= \frac{4\pi}{3}\left[\frac{2c}{H}\left(1 - \frac{1}{\sqrt{3}}\right)\right]^3 n \\
&= \frac{4\pi}{3}\left[\frac{2 \times (3 \times 10^5 \text{ km s}^{-1})}{60 \text{ km s}^{-1} \text{ Mpc}^{-1}}\left(1 - \frac{1}{\sqrt{3}}\right)\right]^3 \times (0.03 \text{ Mpc}^{-3}) \\
&\approx 10^{10} \text{ galaxies.}
\end{aligned}
$$

At one event per day the chances of an event in our Galaxy is 1 in 10^{10} per day or 1 in $10^{10}/365$ per year. Thus in a given galaxy we expect to witness an event every $10^{10}/365 = \boxed{2.7 \times 10^7}$ years.

There must have been several in our Galaxy in the last 500 million years and it is conceivable that a mass extinction event in the Earth's past may have been triggered by such an outburst.

Tutorial 39   The anthropic principle

(a) (i) The mean density of a star of mass M and radius R is $\bar{\rho} = \frac{M}{\frac{4}{3}\pi R^3}$. To the order of approximation used in this question we can ignore factors like $\frac{4}{3}\pi$ so $\bar{\rho} \sim \frac{M}{R^3}$.

(ii) There are several ways to get at the pressure at the centre, but they all amount to the fact that the pressure at any point must be sufficient to hold up the weight of column of material above that point in the star. Pressure is an energy per unit volume (equivalent to a force per unit area), so the pressure will be sufficient if it balances the potential energy per unit volume of the gravitational field, approximately $(GM^2/R)/R^3$. Thus to order of magnitude the pressure P at the centre is

$$P \sim \frac{GM^2}{R^4}. \tag{58}$$

[Somewhat more formally, the equation of hydrostatic equilibrium is

$$\frac{dP}{dr} = -\rho g \tag{59}$$

where the acceleration due to gravity is $g = GM/R^2$ and $\rho \sim M/R^3$ from part (i). The gradient dP/dr can be estimated as P/R. (since the pressure at the surface is negligible compared to that at the centre) giving the result (58) as before. This is equivalent to integrating (59) approximating ρ and g as constants.]

(iii) Taking the material of the Sun to be a perfect gas, the energy per unit volume is nkT where $n = \rho/m_p$ is the number density assuming the material to be pure atomic hydrogen. (Strictly, pure hydrogen is a reasonable asumption, but it is almost completely ionised, not atomic. This means that the partial pressure of free electrons equals that from protons so $P \sim 2nkT$. However, we are not concerned here with factors of 2.) Equating the perfect gas pressure to the expression in (ii) for the pressure required we get

$$kT \sim \frac{GM^2}{R^4} \Big/ \frac{M}{R^3 m_p} = \frac{GM m_p}{R}. \tag{60}$$

(iv) The luminosity is

$$L = \frac{\text{energy density} \times \text{rate of flow}}{\text{area}} = \frac{\left(\frac{aT^4}{\tau}\right) \times c}{4\pi R^2}.$$

The optical depth τ is given by $\sigma n R$, where σ is the cross-section for photons scattering on electrons. This is the Thomson cross-section which is the area of the 'classical' electron πr_e^2. The classical electron radius r_e is given by

$$\frac{e^2}{4\pi\varepsilon_0 r_e} = m_e c^2.$$

We also need the radiation constant a in terms of fundamental constants,

$$a = \frac{k^4}{h^3 c^3}. \tag{61}$$

Putting this together

$$L \sim \left(\frac{k^4}{h^3 c^3}\right) \frac{T^4}{\sigma(\rho/m_p)R} \times c \times \frac{1}{4\pi R^2}$$

$$\sim \left(\frac{G m_p M}{R}\right)^4 \frac{m_p}{h^3 c^2} \frac{1}{\left(\frac{e^2}{4\pi\varepsilon_0 m_e c^2}\right)^2} \frac{1}{\frac{M}{R^3} R^3}$$

$$= \frac{G^4 m_p^5 m_e^2}{\alpha^2 h^5} M^3$$

where $\alpha = e^2/(4\pi\varepsilon_0 hc)$ is the fine structure constant.

(b) If gas pressure exceeds radiation pressure then, as usual to approximate order of magnitude,

$$P_g = nkT > aT^4$$

implies

$$kT < \frac{M^{1/3}}{R} \frac{hc}{m_p^{1/3}},$$

using the explicit form for a given in (61). From (60) this gives

$$M < \left(\frac{hc}{G}\right)^{3/2} \frac{1}{m_p^2}.$$

(c) Assuming a fraction f of a star is consumed in generating the luminosity we get

$$t \sim \frac{fMc^2}{L}.$$

Now putting the results above, taking the upper limit for M from (b), gives

$$t \sim Mc^2 \frac{a^2 h^5}{G^4 m_p^5 m_e^2 M^3} = \frac{a^2 h^2}{cGm_p m_e^2} = \left(\frac{e^2}{4\pi\varepsilon_0}\right)^2 \frac{1}{c^3 Gm_e^2 m_p}.$$

(d) In part (c) we found the dependence of the lifetime on the fundamental constants. This is independent of nuclear reaction rates so it appears that a deuterium composition would make no difference. However, this cannot be so simple since the same argument would show also that the Sun's lifetime would be the same even if there were no nuclear reactions! (Set all the rates to zero and observe the equation is unchanged.) We have clearly assumed that an equilibrium is possible in which nuclear reactions can provide the solar luminosity under the conditions in the centre. We can accept this is true for the actual Sun, so it is true also for a deuterium Sun since increasing the reaction rates makes things easier. Thus the lifetime would indeed be unchanged, but the radius would increase reducing the central temperature and bringing the reaction rates down into line with the flux of radiant energy.

Tutorial 40　Radiating gravity

40.1　Dipole radiation

(a) Compare the inverse square laws for bodies separated by a distance r:

electrostatic force between two charges, e $= \dfrac{1}{4\pi\varepsilon_0}\dfrac{e^2}{r^2}$;

gravitational force between two masses, M $= G\dfrac{M^2}{r^2}$.

Therefore, to convert the Larmor formula to apply to a gravitating mass we make the replacement

$$\frac{e^2}{4\pi\varepsilon_0} \rightarrow GM^2.$$

Thus, the rate of loss of energy of an accelerated mass would be

$$-\frac{dE}{dt} = \frac{2}{3}\frac{GM^2\ddot{x}^2}{c^3}. \tag{62}$$

(b) We need the proportionality between the period T and energy E of a circular orbit.

Kepler's law for an orbit of radius r gives

$$T^2 = \frac{(2\pi)^2}{GM}r^3 \tag{63}$$

The binding energy in a circular orbit is half the gravitational potential energy:

$$E = -\frac{GMm}{2r}. \tag{64}$$

[These results can readily be derived from first principles: Newton's law for a circular orbit is $\frac{mv^2}{r} = \frac{GMm}{r^2}$ from which $v = (GM/r)^{1/2}$. Substituting this into $T = 2\pi r/v$ and using $E = \frac{1}{2}mv^2 - GMm/r$ gives the two results above.]

Thus, eliminating r between (63) and (64) we get

$$T = 2\pi GM \left(\frac{m}{2E}\right)^{3/2}.$$

Finally, by taking logs and differentiating

$$\frac{dT}{dE} = -\frac{3}{2}\frac{T}{E}.$$

For two equal bodies in orbit about a common centre, the separation r is twice the orbital radius a of each. So $r = 2a$, and

$$E = -\frac{GM^2}{4a}.$$

In the Larmor formula (62) we put $x = a\cos\omega t$, and average over a period. The average of $\cos^2 \omega t$ gives a factor $\frac{1}{2}$, and taking into account the radiation from two bodies gives a factor 2. So we estimate

$$L_{\text{larmor}} = -\frac{dE}{dt} = \frac{2}{3}\frac{GM^2}{c^3}\omega^4 a^2.$$

Thus, since $\omega = 2\pi/T$,

$$\frac{-dE/dt}{E} = \frac{8a^3\omega^4}{3c^3} = \frac{128\pi^4 a^3}{3c^3 T^4}.$$

Then, using the data given in the question,

$$\frac{-dE/dt}{E} = \frac{128 \times \pi^4 \times (\frac{2.3}{0.76} \text{ lt sec} \times 3.0 \times 10^8 \text{ m s}^{-1})^3}{3 \times (3.0 \times 10^8 \text{ m s}^{-1})^3 \times (2.8 \times 10^4 \text{ s})^4} = 1.9 \times 10^{-13} \text{ s}^{-1}$$

and

$$\frac{dT}{dt} = \frac{-dE/dt}{E} \times E\frac{dT}{dE} = \frac{-dE/dt}{E} \times \frac{-3T}{2}$$

$$= -\frac{3}{2} \times (1.9 \times 10^{-13} \text{ s}^{-1}) \times (2.8 \times 10^4 \text{ s})$$

$$= -8.0 \times 10^{-9} \text{ s s}^{-1}$$

which is a factor 3000 more than the observed value of 2.7 x 10^{-12}.

40.2   Quadrupole radiation

(a) We are given that $L_{\text{grav}} = K\left\langle \overset{...}{I}^2 \right\rangle$. The dimensions of the luminosity are

$$[L_{\text{grav}}] = [\text{energy per unit time}] = ML^2T^{-2} \times T^{-1} = ML^2T^{-3}.$$

The dimensions of $\overset{...}{I}^2$ are

$$[\overset{...}{I}^2] = (ML^2 \times T^{-3})^2 = M^2L^4T^{-6}.$$

The constant of proportionality K must therefore have the dimensionality $M^{-1}L^{-2}T^{-3}$. Since the formula arises in the context of relativistic gravity (the general theory of relativity) the constant can involve only c and G. We have

$$[c] = LT^{-1} \text{ and } [G] = [\text{Force}] \times M^{-2}L^2 = M^{-1}L^3T^{-2}.$$

Therefore

$$K = \frac{c^5}{G}.$$

In fact, $L_{\text{grav}} = (128/5) \times (c^5/G)\left\langle \overset{...}{I}^2 \right\rangle$.

For the binary pulsar (mass M, semi-major axis a) we estimate $I \simeq 2Ma^2$, so for a period T, we have $\left\langle \overset{...}{I}^2 \right\rangle \simeq (2Ma^2/T^3)^2 \simeq (2Ma^2\omega^3)^2$, omitting factors of 2π. Denoting the radiation power from the Larmor formula (62)as L_{larmor} our dimensional analysis gives

$$\frac{L_{\text{larmor}}}{L_{\text{grav}}} \simeq \left(\frac{2GM^2a^2\omega^4}{3c^3}\right)\left(\frac{c^5}{4GM^2a^4\omega^6}\right)$$

$$= \frac{c^2}{6a^2\omega^2}$$

$$= \frac{(3 \times 10^8)^2}{6 \times \left(\frac{2.34}{0.76} \times 3 \times 10^8\right)^2\left(\frac{2\pi}{2.8\times 10^4}\right)^2} \approx 3 \times 10^5.$$

Thus the gravitational radiation is many orders of magnitude less than predicted by applying the analogue Larmor formula.

To explain the discrepancy between the Larmor formula and observation in part (b) requires a factor $\approx 3 \times 10^3$. The fact that we have obtained a factor 3×10^5

does not imply that general relativity is also wrong: our order of magnitude estimate leaves out some large numerical factors, for example 128/5 in L_{grav}. In fact, when the calculation is done properly, the observations on the decay of the orbit are in accordance with the predictions from relativity of the energy radiated. This means there is no room for any significant increase in the loss of energy by radiation by other mechanisms, which rules out a number of various alternative theories to general relativity that have been proposed.

(b) We are given that $dT/dt = -2.40 \times 10^{-12}$. From this we estimate the power in gravitational radiation using the results of question (a). We have

$$-\frac{dE}{dt} = -\frac{dE}{dT}\frac{dT}{dt} = -\frac{2}{3}\frac{E}{T}\left(\frac{dT}{dt}\right) = \frac{2}{3}\left(\frac{GMm}{4aT}\right)\left(-\frac{dT}{dt}\right)$$

$$\approx \frac{2 \times (6.7 \times 10^{-11}\mathrm{N\,m^2\,kg^{-2}}) \times (1.4 \times 2.0 \times 10^{30}\mathrm{kg})^2}{12 \times \left(\frac{2.34}{0.76}\mathrm{lt\ s} \times 3 \times 10^8\mathrm{m\,s^{-1}}\right) \times (2.8 \times 10^4\mathrm{s})} \times 2.4 \times 10^{-12}$$

$$\approx 8 \times 10^{24}\ \mathrm{W}.$$

The flux at the Earth is

$$\frac{-dE/dt}{4\pi d^2} = \frac{8 \times 10^{24}\ \mathrm{W}}{4\pi \times (5 \times 3.1 \times 10^{19}\ \mathrm{m})^2} = \boxed{2.6 \times 10^{-17}\ \mathrm{W\,m^{-2}}.}$$

For comparison, it is hoped that the best gravitational wave detectors might be able to detect supernovae in the Galaxy producing 10^{47} W in a millisecond burst.

Tutorial 41   Radioactive decay

41.1   A natural fission reactor

The decay of ^{235}U is given by

$$^{235}N = {}^{235}N_0 \exp(-\lambda_5 t)$$

and the decay of ^{238}U similarly by

$$^{238}N = {}^{238}N_0 \exp(-\lambda_8 t).$$

Thus, dividing one equation by the other gives the decay rate for the ratio:

$$\frac{^{235}N}{^{238}N} = \frac{^{235}N_0}{^{238}N_0}e^{-(\lambda_5-\lambda_8)t}. \tag{65}$$

Take the initial enrichment of ^{235}U to be the minimum required for a chain reaction (3%) to obtain a lower limit to the age. Thus $^{235}N_0/^{238}N_0 = 0.03$.

At the present time $^{235}N/^{238}N = 0.0072$ so the time required is given by

$$e^{(\lambda_5-\lambda_8)t} = \frac{0.03}{0.0072},$$

or

$$t = \frac{\log_e \frac{3}{0.72}}{\left(\frac{\log_e 2}{7.13 \times 10^8} - \frac{\log_e 2}{4.51 \times 10^9}\right)} = \boxed{1.74 \times 10^9 \text{ yrs}}.$$

The Oklo reactors must have operated earlier than 1.74×10^9 years ago, but they could not predate the time when liquid water first formed on the Earth.

41.2   The Uranium clock

Using equation (65) and setting $^{235}N_0 = {}^{238}N_0$ we have
$$e^{-(\lambda_5 - \lambda_8)t} \simeq 0.0072,$$

from which $\boxed{t = 6 \times 10^9 \text{ years}}$. The mean time of formation of the heavy elements on Earth was therefore about 1.5 billion years before the formation of the solar system.

41.3   Natural Pu

Uranium undergoes spontaneous fission releasing on average about 2.5 neutrons per fission. Each neutron absorbed by a ^{238}U nucleus gives a ^{239}U nucleus,
$$^{238}U + n \rightarrow {}^{239}U.$$

which undergoes two beta decays to ^{239}Pu. The concentration of Pu will come into secular equilibrium with the much longer lived ^{238}U. If the decay constants for ^{239}Pu and ^{238}U are λ_9 and λ_8 repectively, and each decay of ^{238}U produces 2.5 neutrons, this implies
$$\lambda_9 N_9 = 2.5 \lambda_8 N_8.$$

Thus,
$$N_9 = 2.5 \frac{[T_{1/2}]_9}{[T_{1/2}]_8} \times N_8 = 2.5 \times \frac{24.4 \times 10^3}{10^{16}} N_8 = 6.1 \times 10^{-12} N_8.$$

Therefore 1kg of ^{238}U will contain about 6.1×10^{-12} kg of ^{239}Pu, or about $(6.1 \times 10^{-12} \text{ kg})/[239 \times (1.67 \times 10^{-27} \text{ kg})] \approx \boxed{1.5 \times 10^{13} \text{ atoms}}$.

Tutorial 42   Elementary particles

42.1   Catching neutrinos

(a) The maximum number of Ar atoms will be attained at equilibrium when the rate of creation equals the rate of decay. The rate of creation is given here to be $R = 0.5$ atoms per day. Thus, if λ is the rate constant for the radioactive decay of ^{37}Ar, the equilibrium condition is $\lambda N_{eq} = R$, and the equilibrium abundance of ^{37}Ar is

$$N_{eq} = \frac{R}{\lambda} = \frac{0.5 \text{ day}^{-1}}{\log_e 2/(37 \text{ days})} = 26.7 \text{ atoms}$$

or $\boxed{N_{eq} \approx 27 \text{ atoms}}$.

(b) To obtain the number of atoms $N(t)$ present after a given time t we solve the rate equation

$$\frac{dN}{dt} = R - \lambda N.$$

This differential equation has the solution

$$N = \frac{R}{\lambda} \left(1 - e^{-\lambda t}\right) = N_{eq} \left(1 - e^{-\lambda t}\right).$$

So after 3 months

$$N = 27 \times \left(1 - \exp\left[-\left(\frac{\log_e 2}{37 \text{ days}}\right) \times (90 \text{ days})\right]\right) \approx \boxed{22 \text{ atoms.}}$$

42.2   The long-lived proton

(a) The energy absorbed by the body depends on the energy released by each proton decay, the number of protons in the body and the decay rate. We treat each in turn.

The decay of a proton releases its rest mass energy of $E = 938$ MeV which we assume is completely absorbed in the body.

The number, N, of protons per kg of water is given by

$$N = \begin{array}{c} \text{No. of water} \\ \text{molecules per mole} \end{array} \times \begin{array}{c} \text{No. of moles} \\ \text{per kg} \end{array} \times \begin{array}{c} \text{No. of protons} \\ \text{per water molecule} \end{array}$$

$$= (6 \times 10^{23}) \times \left(\frac{1000}{18}\right) \times 10$$

$$= 33 \times 10^{25} \text{ kg}^{-1}.$$

Let the decay rate be λ per year. Then the number of decays per year per kg is $N\lambda$ and the energy per year per kg is $EN\lambda$. This must equal 50×10^{-3} Sv or 0.05 J kg^{-1} yr^{-1}. Thus the mean lifetime $1/\lambda$ is given by

$$\frac{1}{\lambda} = \frac{EN}{0.05} = \frac{(938 \times 10^6 \text{ eV}) \times (1.6 \times 10^{-19} \text{ J eV}^{-1}) \times (33 \times 10^{25} \text{ kg}^{-1})}{0.05}$$

$$= \boxed{9.9 \times 10^{17} \text{ yr.}}$$

(b) To detect 1 decay per year we need a number of atoms N such that $N\lambda = 1$ i.e. we need 10^{33} protons. From part (a) 1 kg contains 33×10^{25} protons, so the mass of water required is

$$\frac{10^{33} \text{ protons}}{33 \times 10^{25} \text{ protons kg}^{-1}} \approx 3 \times 10^6 \text{ kg}$$

or about $\boxed{3000 \text{ tonnes}}$.

Tutorial 43   Quantum uncertainty

43.1   An alternative Bohr

(a) The electron has a total energy

$$E = \frac{p^2}{2m} - \frac{Ze^2}{4\pi\epsilon_0 r}.$$

If the electron is confined within a radial distance r then according to the uncertainty principle it must have a momentum of at least $p \sim \hbar/r$. The minimum allowable kinetic energy is therefore

$$\frac{p^2}{2m} = \frac{\hbar^2}{2mr^2}.$$

Thus, for hydrogen ($Z = 1$)

$$E = \frac{\hbar^2}{2mr^2} - \frac{e^2}{4\pi\epsilon_0 r}. \tag{66}$$

(b) We obtain the radius of a stable configuration by minimising the energy as usual:

$$\frac{dE}{dr} = \frac{-2\hbar^2}{2mr^3} + \frac{e^2}{4\pi\epsilon_0 r^2} = 0$$

giving

$$r_{\min} = \frac{4\pi\epsilon_0\hbar^2}{me^2} = 5.26 \times 10^{-11} \text{ m} = \boxed{0.53 \text{ Å}}.$$

(c) This is identical to the result from the Bohr theory. The minimum (ground state) energy is

$$E_{\min} = \frac{\hbar^2 m^2 e^4}{2m(4\pi\epsilon_0)^2\hbar^4} - \frac{e^2 \times me^2}{(4\pi\epsilon_0)^2\hbar^2} = -\frac{me^4}{8\epsilon_0^2 h^2}.$$

43.2   Pressure ionisation

(a) The electron becomes unbound if the radius r of its orbit is reduced to the point where the energy $E = 0$. From (66) this happens when

$$r = r_{\text{ion}} = \frac{4\pi\epsilon_0\hbar^2}{2me^2} = \boxed{\tfrac{1}{2}r_{\min}}. \tag{67}$$

(b) The pressure required to achieve this can be obtained using

$$P = -\frac{\partial E}{\partial V} = -\frac{\partial E}{\partial r}\frac{dr}{dV} = -\frac{\partial E}{\partial r}\frac{1}{4\pi r^2}$$

evaluated at $r = r_{ion} = \frac{1}{2}r_{min}$ (where $V = \frac{4}{3}\pi r^3$). This gives

$$P = -\frac{1}{4\pi r^4}\left(\frac{-\hbar^2}{mr} + \frac{e^2}{4\pi\varepsilon_0}\right),$$

from which

$$
\begin{aligned}
P_{ion} &= \frac{1}{4\pi r_{ion}^4}\left(\frac{\hbar^2}{m} \times \frac{2me^2}{4\pi\varepsilon_0\hbar^2} - \frac{e^2}{4\pi\varepsilon_0}\right) \\
&= \frac{1}{4\pi r_{ion}^4}\left(\frac{e^2}{4\pi\varepsilon_0}\right) \\
&= \frac{(1.6 \times 10^{-19}\ \text{C})^2}{16\pi^2(8.85 \times 10^{-12}\ \text{F m}^{-1}) \times (2.63 \times 10^{-11}\ \text{m})^4} \\
&= \boxed{3.8 \times 10^{13}\ \text{N m}^{-2}.}
\end{aligned}
$$

(c) According to the definition a planet has the maximum mass M when the pressure at the centre is just equal to P_{ion}, the pressure beyond which matter becomes ionised. This pressure at the centre must balance gravity by supporting a column of material of height equal to the radius R of the planet. If the average density is ρ then, to order of magnitude, the pressure must be $\rho g R$, where g, the acceleration due to gravity, is of order GM/R^2. Thus the pressure must be of order $GM\rho/R$.

Note that it is not possible to obtain a more accurate value without an equation of state to relate P and ρ throughout the interior. Given $P(\rho)$ the pressure distribution would be obtained by integrating the equation of hydrostatic equilibrium, $dP/dr = -GM(r)\rho/r^2$.

Now, equating this pressure to P_{ion}, we obtain

$$\frac{1}{4\pi r_{ion}^4}\frac{e^2}{4\pi\varepsilon_0} \sim \frac{GM\rho}{R}. \tag{68}$$

But, if the atomic size is r_{ion} as a result of compression, we can estimate the average density as

$$\rho \sim \frac{m_p}{\frac{4}{3}\pi r_{ion}^3} \tag{69}$$

Also, $M \sim \frac{4}{3}\pi R^3\rho$, from which,

$$R = \left(\frac{3M}{4\pi\rho}\right)^{1/3}. \tag{70}$$

Using (67) for $r_{\rm ion}$ and (69) and (70) for ρ and R in (68) we obtain

$$M = \left(\frac{3^{1/3}e^2}{4\pi\varepsilon_0 m_p^{4/3}G}\right)^{3/2}$$

$$= \frac{\sqrt{3}\times(1.6\times10^{-19}\,{\rm C})^3}{[4\pi(8.85\times10^{-12}{\rm F\,m^{-1}})\times(6.67\times10^{-11}{\rm N\,m^2kg^{-2}})]^{3/2}\times(1.7\times10^{-27}{\rm kg})^2}$$

$$\approx \boxed{4\times10^{27}\ {\rm kg.}}$$

This is sometimes called the Fowler mass. The mass of Jupiter is 1.9×10^{27} kg and its density is 1330 kg m^{-3}.

43.3   Nuclei without neutrons

(a) The atomic weight of gold is $A = 197$, so the radius of the gold nucleus is
$$r = (1.2\times10^{-15}\ {\rm m})\times(197)^{1/3} = 7\times10^{-15}\ {\rm m}.$$
For an electron of momentum p confined to a radius r the uncertainty principle implies $p\sim\hbar/r$.

The kinetic energy of the electron confined to the nucleus would therefore be

$$\begin{aligned} p^2/2m_e &\sim \frac{(1.05\times10^{-34}\ {\rm J\,s})^2}{(7\times10^{-15}\ {\rm m})^2\times2\times(9.11\times10^{-31}\ {\rm kg})}\\ &= 1.2\times10^{-10}\ {\rm J}\\ &= \boxed{7.7\times10^8\ {\rm eV}}. \end{aligned}$$

(b) The atomic number of gold is 79 so the electrostatic energy at the surface of the nucleus is
$$\begin{aligned} E_{\rm es} &= \frac{Ze^2}{4\pi\varepsilon_0 r} = \frac{79\times(1.6\times10^{-19}C)^2}{4\pi\times(8.8\times10^{-12}Fm^{-1})\times(7\times10^{-15}m)}\\ &= 2.6\times10^{-12}\ {\rm J}\\ &= \boxed{1.6\times10^7\ {\rm eV.}} \end{aligned}$$
The kinetic energy of the electron in the nucleus is therefore much greater than the electrostatic potential energy, so the electron will escape.

(c) For a neutron (or a proton) the corresponding kinetic energy is of order
$$p^2/2m_p = (7.7\times10^8\ {\rm eV})\times(m_e/m_p) = \frac{7.7\times10^8}{1837}\ {\rm eV} = 4.2\times10^5\ {\rm eV},$$
which is much less than the value we found for the electron. On the other hand the nuclear force is much stronger than the electrostatic force so the binding energy of the neutron (or proton) in a nucleus is much greater than $E_{\rm es}$. Therefore

the neutrons and protons are bound.

Tutorial 44   Cross sections

44.1   Supernova neutrinos

(a) Assuming a stellar core of uniform density and mass M, the gravitational binding energy released in collapse from radius r_i to radius r_f is

$$\text{B.E.} = \frac{3}{5}GM^2 \left(\frac{1}{r_f} - \frac{1}{r_i} \right).$$

(This is derived from the potential energy $\int \frac{GMdM}{r}$ for a uniform sphere. Since the true density distribution is not uniform the expression is only an estimate, so the factor 3/5 which comes from the exact integration could be omitted.) Since $r_i \gg r_f$ the binding energy is approximately $(3/5) \times (GM^2/r_f)$. So

$$\text{B.E.} \approx \frac{3}{5} \times \frac{(6.67 \times 10^{-11} \text{ N kg}^{-2} \text{ m}^2) \times (1.4 \times 2 \times 10^{30} \text{ kg})^2}{10^4 \text{ m}} = \boxed{3.14 \times 10^{46} \text{ J.}}$$

(b) Since N/A is the time integrated flux at the detector and the number of proton targets $= nV$, where the volume $V = A\Delta x$, the result follows:

$$\begin{matrix} \text{Number of} \\ \text{neutrino} \\ \text{captures} \end{matrix} = \text{Cross section} \times \begin{matrix} \text{Time integrated flux} \\ \text{of } \nu_e \text{ at the detector} \end{matrix} \times \begin{matrix} \text{Number of protons} \\ \text{in the detector} \end{matrix}$$

or

$$\Delta N = \sigma \times \frac{N}{A} \times nA\Delta x \tag{71}$$

(c) We use this relation to obtain the flux and show that the total energy in neutrinos is consistent with the energy released in collapse.

$$\begin{aligned} \text{Number of free protons} \ &= \ 2 \times \text{ number of water molecules} \\ &= \ 2 \times \text{ number of moles } \times N_0 \\ &= \ 2 \times \frac{(2140 \times 10^3 \times 10^3 g)}{(18 \text{ g mole}^{-1})} \times (6 \times 10^{23}) \\ &= \ 1.42 \times 10^{32}. \end{aligned}$$

Now, using (71) we get

Time integrated flux

$$= \frac{11 \text{ events}}{(1.42 \times 10^{32} \text{ protons}) \times (9.23 \times 10^{-46} \text{ m}^2) \times (14 \text{ MeV}/10 \text{ MeV})^2}$$
$$= \ 4.3 \times 10^{13} \text{ m}^{-2}.$$

When account is taken of the known detector efficiency as a function of energy the integrated flux comes to 10^{14} m^{-2}.

Assuming the neutrinos are emitted isotropically, then

$$
\begin{aligned}
\text{Total } \nu_e &= \text{Integrated flux} \times 4\pi \times (\text{distance of source})^2 \\
&= (4.3 \times 10^{13} \text{ m}^{-2}) \times 4\pi \times (5 \times 10^4 \text{ pc} \times 3.086 \times 10^{16} \text{ m pc}^{-1})^2 \\
&= 1.3 \times 10^{57}
\end{aligned}
$$

The total energy of these ν_e is
$(1.4 \times 10^7 \text{ eV}) \times (1.6 \times 10^{-19} \text{ J eV}^{-1}) \times (1.3 \times 10^{57} \text{ neutrinos}) = 2.9 \times 10^{45} \text{ J}$,
and multiplying by 6 to take account of the 3 neutrino flavours each with an anti-neutrino partner, the total energy is $\boxed{1.8 \times 10^{46} \text{ J}}$.

If the detector efficiency is taken into account the result is an energy of 4×10^{46} J. This is close to the estimated gravitational binding energy. (*Phys Rev Lett*, 1987, **58**, 1490).

44.2   Yukawa's meson

(a) The equation of motion of the muon, in a circular orbit of radius r_μ with speed v in an atom of atomic number Z is

$$
\frac{mv^2}{r} = \frac{Ze^2}{4\pi\varepsilon_0 r^2}.
$$

Bohr's postulate for the quantisation of angular momentum gives

$$
mvr_\mu = n\hbar.
$$

Eliminating v yields

$$
r_\mu = \frac{4\pi\varepsilon_0 n^2 \hbar^2}{Ze^2 m}.
$$

Thus, $r_\mu/r_e = m_e/m_\mu = 1/207$.

(b) The result of (a) shows that in its ground state the muon orbit lies far within the electron orbits. Since the field inside a shell of charge is zero, the field of the electrons can be ignored, and only the electrostatic field of the nucleus is important.

(c) From (a)

$$
r_\mu = \frac{4\pi\varepsilon_0 (h/2\pi)^2}{m_\mu Ze^2} = \frac{4\pi \times (8.8 \times 10^{-12} \text{ F m}^{-1}) \times (6.6 \times 10^{-34} \text{ J s}/2\pi)^2}{(207 \times 9.1 \times 10^{-31} \text{ kg}) \times 82 \times (1.6 \times 10^{-19} \text{ C})^2}
$$

$$
= 3.1 \times 10^{-15} \text{ m}.
$$

The radius of the lead nucleus is $r_N = (1.2 \times 10^{-15} \text{m}) \times (208)^{1/3} = 6.5 \times 10^{-15}$m, from which we deduce that the muon lies inside the nucleus. In fact, this is only approximately correct: since the muon is inside the nucleus we should not treat the nucleus as a point charge. However, a full quantum mechanical treatment shows that the finite size of the nucleus means that the muon spends about 50% of the time within the nucleus. Our approximate treatment is therefore reasonable.

(d) Consider a muon moving through nuclear matter with speed $v \sim c$. If the number density of nucleons is n, the mean free time is

$$\bar{t} = \frac{1}{n\sigma c}.$$

Within the nucleus, the number density of nucleons is

$$n = \frac{\text{number of nucleons}}{\text{volume of nucleus}} = \frac{A}{\frac{4}{3}\pi(1.2 \times 10^{-15}A^{1/3} \text{ m})^3} = 1.8 \times 10^{44} \text{ m}^{-3}.$$

Therefore,

$$\sigma = \frac{1}{n\bar{t}c} = \frac{1}{(1.8 \times 10^{44} \text{ m}^{-3}) \times (10^{-8} \text{ s}) \times (3 \times 10^8 \text{ m s}^{-1})} \approx \boxed{10^{-45} \text{ m}^2}.$$

This value for a cross section is typical for a weak interaction process. Therefore the muon does not interact strongly with nucleons and so cannot be Yukawa's meson.

(e) The 'something' in the reaction equation has to be a muon neutrino to conserve muon lepton number.

44.3 Too far flung for civilisation

We want to calculate the mean free time between close stellar encounters, where by 'close' we mean close enough to perturb seriously any planetary orbits about a star. If this timescale were short by biological evolutionary standards life would presumeably not emerge. The mean free time for collisions between stars moving with speed v in an environment with stellar density n is

$$\bar{t} = \frac{1}{n\sigma v},$$

where σ is the collision cross section. We know n and v but we need to estimate σ. A star passing within the orbit of Jupiter, a distance of $r_J = 7.8 \times 10^8$ km would certainly affect the Earth, so take $\sigma = \pi r_J^2$. This probably overestimates $\bar{t}$. We get

$$\bar{t}^{-1} = \left(\frac{10^6 \text{ pc}^{-3}}{(3.1 \times 10^{16} \text{ m pc}^{-1})^3}\right) \times \pi(7.8 \times 10^{11} \text{ m})^2 \times (2 \times 10^5 \text{ m s}^{-1}) \times$$
$$\times (3.2 \times 10^7 \text{ s yr}^{-1})$$

or $\bar{t} = 2.4 \times 10^6$ yr.

This is too short a time for the development of life, and certainly too short for the development of advanced civilisation. (On Earth the fossil record shows the process took at least 3.5 billion years.)

This is another of our favourite questions - one seems to get back such a great deal of information from such a small amount of data.

Tutorial 45   Nuclear explosions

45.1   Critical mass

(a) The existence of a critical mass below which a self-sustaining chain reaction is not possible is an example of the 'square-cube law'. The fraction of fission neutrons which escape across the surface of a lump of fissionable material can be reduced by increasing the size of the lump since the rate of production is proportional to the volume (R^3) whilst the rate of escape is proportional to the area (R^2). At the critical mass on average each fission gives rise to an extra fission which is the condition for a self-sustaining chain reaction. A spherical lump minimises the critical mass as a sphere minimises the surface area to volume ratio.

(b) The mean free path for neutrons in a medium of number density n of target nuclei with capture cross-section σ is $\lambda = 1/(\sigma n)$. Thus $\lambda \propto 1/\rho$ in a medium of mass density ρ.

Now suppose the mean free path changes for some reason. How does the radius R_c of the critical mass change? Clearly R_c depends on λ and since there is no other relevant length scale we must have $R_c \propto \lambda$. (The same result can be obtained by balancing production and loss rates of neutrons.)

So $M_c = \frac{4}{3}\pi\rho R_c^3 \propto \rho\lambda^3 \propto 1/\rho^2$.

(c) From off-centre points the probability of escape from the nearer surface is increased and decreased for the further surface. To order of magnitude it is therefore constant through the bulk of the material, with the exception of atoms near the surface, which, however, make up a negligible fraction of the material. For a sphere of radius r and number density of ^{235}U atoms n the number of neutrons Q escaping from the surface is related to the number produced at the centre Q_0 by

$$Q = Q_0 e^{-n\sigma r}.$$

For a critical mass

$$1.5 = 2.5 e^{-n\sigma R_c}.$$

Thus $R_c = \frac{1}{n\sigma}\log_e\left(\frac{2.5}{1.5}\right) = \frac{0.51}{n\sigma}$. We obtain n from the given density: $n =$

$\rho/(\text{mass of an atom}) = (\rho N_A)/0.235$. Thus

$$M_c = \frac{4}{3}\pi R_c^3 \rho$$

$$= \frac{4\pi}{3}\left(\frac{0.51 \times 0.235 \text{ kg}}{(6 \times 10^{23}\text{mole}^{-1}) \times (1.22 \times 10^{-28}\text{m}^2)}\right)^3 \frac{1}{(19000 \text{ kg m}^{-3})^2}$$

$$= \boxed{51.2 \text{ kg.}}$$

This, suprisingly for such a rough estimate, is not far from the value of 49kg for weapons grade uranium (93.5% ^{235}U) (*Nature*, **238**, 817, 1980).

45.2   Nuclear fireballs

(a) 1kg of ^{239}Pu contains $N = \frac{1000}{239} \times 6 \times 10^{23} = 2.51 \times 10^{24}$ atoms.
The energy released from fission of 50% of these atoms is

$$E = \frac{1}{2} \times (2.51 \times 10^{24}\text{atoms}) \times (180 \times 10^6\text{eV}) \times (1.6 \times 10^{-19} \text{ J eV}^{-1})$$

$$= \frac{1}{2} \times 7.23 \times 10^{13} \text{ J} = \boxed{3.6 \times 10^{13} \text{ J}}.$$

(b) The atoms of plutonium will be completely ionised by this energy release. (Compare the ionisation energy of a few keV with 180 MeV.) So the energy is divided between the nuclei, the 94 free electrons per nucleus and radiation in thermal equilibrium with these particles.

If the volume occupied by the 2kg mass of plutonium is V the total energy $2E$ is

$$2E = \left(\frac{96 \text{ particles}}{\text{per fission}}\right) \times (2N \text{ particles in 2 kg}) \times (\frac{3}{2}kT \text{ per particle}) + VaT^4$$

where the first term is the particle energy in 2kg and the second is the energy in radiation. Therefore

$$7.23 \times 10^{13} = 3 \times 96 \times (2.51 \times 10^{24}\text{nuclei kg}^{-1}) \times (1.38 \times 10^{-23}\text{J K}^{-1}) \times T$$

$$+\frac{(2 \text{ kg})}{(1.9 \times 10^4 \text{ kgm}^{-3})}(7.56 \times 10^{-16}\text{J K}^{-4})T^4$$

or

$$7.23 \times 10^{13} = 9.87 \times 10^3 T + 7.96 \times 10^{-20}T^4$$

with T in Kelvin. To solve this equation first check if one term dominates. If the first term on the right dominates then $T \sim 10^{10}$K for which value the second term is in fact the larger. If the second term dominates $T \sim 10^8$K for which value the first term is negligible. Thus, it is reasonable to neglect the matter energy, in which case we obtain $\boxed{T = 1.7 \times 10^8\text{K}}$.

(c) Since pressure is of order the energy density the main contribution to the pressure from part (b) is the radiation. The radiation pressure is

$$\frac{1}{3}aT^4 = \frac{1}{3} \times (7.56 \times 10^{-16} \text{J m}^{-3} \text{ K}^{-4})(1.7 \times 10^8 \text{K})^4$$

$$= 2.1 \times 10^{17} \text{ N m}^{-2} = \boxed{2.1 \times 10^{12} \text{ atmospheres.}}$$

(d) From Wien's displacement law $\lambda T = 2898 \times 10^{-6}$ mK from which $\lambda = 0.29$Å. Equivalently,

$$h\nu = hc/\lambda = \frac{(6.67 \times 10^{-34} \text{J s}) \times (3 \times 10^8 \text{m s}^{-1})}{(0.29 \times 10^{-10}\text{m}) \times (1.6 \times 10^{-19}\text{J eV}^{-1})} = \boxed{42.3 \text{ keV}}.$$

The realization that most of the energy is in γ-rays was a major step in the development of the Pu bomb as a trigger device for the H bomb. Gamma rays from the fission bomb compress and heat the D-T mixture causing it to fuse before the device is blown apart.

(e) The γ-rays ionise air atoms. This rapid separation of charges produces an electromagnetic pulse.

Tutorial 46   Degenerate electron gases

46.1   A matter of compressibility

(a) At $T = 0$ we have

$$dU = -PdV.$$

In terms of the fixed total number of electrons $N = nV$, the energy density is

$$U = AN^{5/3}V^{-2/3}, \tag{72}$$

where

$$A = \frac{4\pi h^2}{5m}\left(\frac{3}{8\pi}\right)^{5/3} = \frac{4\pi}{5} \times \frac{(6.6 \times 10^{-34}\text{Js})^2}{(9.1 \times 10^{-31}\text{kg})} \times \left(\frac{3}{8\pi}\right)^{5/3} = 3.48 \times 10^{-38} \text{ J}^2 \text{ s}^2 \text{ kg}^{-1}$$

Therefore,

$$P = -\frac{dU}{dV} = \frac{2}{3}AN^{5/3}V^{-5/3} = \frac{2}{3}An^{5/3}.$$

Reducing the volume increases the kinetic energy of the electrons, so they exert greater a pressure.

(b) Inserting values into the expression for pressure

$$P = \frac{2}{3} \times (3.48 \times 10^{-38} \text{ J}^2 \text{ s}^2 \text{ kg}^{-1}) \times (1.1 \times 10^{29} \text{ m}^{-3})^{5/3} = \boxed{5.86 \times 10^{10} \text{ Pa}}.$$

The bulk modulus is

$$B = -V\frac{dP}{dV} = \frac{5}{3}P \approx 10^{11} \text{ Pa.}$$

This estimate leaves out the effect of the ion cores. Nevertheless it gives a sensible value, showing that the exclusion principle is principally responsible for the incompressibility of metals. Some experimental values are 1.4×10^{11} Pa for copper and 0.07×10^{11} Pa for lead.

(c) The work done in compressing a volume V_0 of electron gas by a factor 2 is (from (72))

$$\Delta U = AN^{5/3}V_0^{-2/3}(2^{2/3} - 1).$$

where

$$N = \frac{(2000 \text{ gm})}{(239 \text{ gm mole}^{-1})} \times 6 \times 10^{23} \text{mole}^{-1} = 5.02 \times 10^{24}.$$

The volume of a 2 kg plutonium sphere is

$$(2 \text{ kg})/(1.9 \times 10^4 \text{ kg m}^{-3}) = 1.05 \times 10^{-4} \text{ m}^{-3},$$

so

$$\Delta U = (3.48 \times 10^{-38} \text{ J}^2 \text{ s}^2 \text{ kg}^{-1}) \times (5.02 \times 10^{24})^{5/3} \times \frac{(2^{2/3} - 1)}{(1.05 \times 10^{-4}\text{m}^3)^{2/3}}$$

$$= 1.35 \times 10^6 \text{ J}.$$

The energy yeild from a kg of high explosives is several times this amount. (See tutorial 1.) Not all the explosive energy released will go into compressing the plutonium, but 1kg of high explosive should be sufficient to achieve a factor 2 compression.

46.2 Brown dwarfs

The total energy of a brown dwarf of mass M, radius R with internal energy U is

$$E = U - f\frac{GM^2}{R},$$

where f is a numerical factor that depends on the distribution of density with radius in the star. For a star of uniform density we should have $f = 3/5$. Expressing this in terms of the total number of electrons $N = \frac{1}{2}M/m_p$, (since we are given that the number of nucleons is twice the number of electrons) and the volume $V = \frac{4}{3}\pi R^3$, we have

$$E \approx AN^{5/3}V^{-2/3} - 4DN^2V^{-1/3},$$

where A is defined above and

$$D = f(\frac{3}{4\pi})^{-1/3}Gm_p^2$$

$$= f(\frac{3}{4\pi})^{-1/3} \times (6.67 \times 10^{-11} \text{ N m}^2\text{kg}^{-2}) \times (1.6 \times 10^{-27} \text{ kg})^2$$

$$= 2.75 \times 10^{-64} f \text{ N m}^2.$$

The equilibrium configuration will occur at the minimum of the total energy, $dE/dV = 0$. This condition is

$$-\frac{2}{3}AN^{5/3}V^{-5/3} + \frac{4}{3}DN^2V^{-4/3} = 0$$

from which $V \propto N^{-1}$ or $R \propto M^{-1/3}$. The behaviour of this material is somewhat counter-intuitive. The largest brown dwarfs are smaller than Jupiter in size despite being 84 times more massive.

46.3   White dwarfs

The expression for the total energy of self-gravitating relativistic degenerate matter is

$$E = 2\pi chV \left(\frac{3}{8\pi}\right)^{4/3} N^{4/3}V^{-4/3} - f\frac{G(2Nm_p)^2}{\left(\frac{3V}{4\pi}\right)^{1/3}}$$

$$= CN^{4/3}V^{-1/3} - 4DN^2V^{-1/3},$$

where $C = 2\pi ch \left(\frac{3}{8\pi}\right)^{4/3} = 2\pi \times (3\times 10^8 \text{ m s}^{-1}) \times (6.62\times 10^{-34} \text{ J s})\times\left(\frac{3}{8\pi}\right)^{4/3} = 7.3 \times 10^{-26}$ J m is another constant. Both terms have the same V dependence so there is no state that minimises the energy. If $E > 0$ the system will expand thus lowering the Fermi energy below mc^2 so the material ceases to be completely relativistic and the star can find a stable state. If $E < 0$ the system will collapse with nothing to halt it (until the material becomes neutron matter). Thus, $E = 0$ divides stable and unstable systems. This occurs for N given by

$$N = \left(\frac{C}{4D}\right)^{3/2} = \left(\frac{7.3 \times 10^{-26} \text{ J m}}{4 \times 2.75 \times 10^{-64} f \text{ N m}^2}\right)^{3/2} = 5.41 \times 10^{56} f^{-3/2}.$$

For a uniform sphere ($f = 3/5$) this gives a mass $M = 2Nm_p = 2 \times 5.41 \times 10^{56} \times (3/5)^{-3/2} \times (1.6 \times 10^{-27} \text{ kg}) = 3.73 \times 10^{30} \text{ kg or } \boxed{1.87M_\odot}$.

A detailed calculation gives $1.44M_\odot$.

Part 3: General Physics Problems

1   Energetics

1) The total consumption of energy in the UK is about 10^{19} J yr^{-1}, about 1/30 th of the world total. What fraction of this is required for domestic heating? Assume that the rate of heat loss through a domestic cavity wall is about 2 W m^{-2} per unit centigrade temperature difference across the wall.

2) (a) Estimate the Lorentz γ-factor of a spaceship which travels to a nearby star 10 pc from the solar system in 50 years of spaceship time.

(b) Put a lower bound on the energy that such a trip would require by considering just the kinetic energy of the payload.

(c) In order to put this amount of energy into an everyday context estimate the annual world consumption of energy given that the world oil production is currently 25 billion barrels per year (*Nature*, 8 May, 1997). One barrel of oil is 0.159 m^3.

3) The estimated world reserves of oil are put at about 1000 billion barrels. Would these reserves supply enough energy to accelerate the interstellar spaceship to its cruising speed?

2   Mechanics in action

Astronauts need experience of weightless conditions before they go into orbit. These conditions can be realised for a short time inside an aircraft that is following a free-fall trajectory. This method was used for the weightless scenes in the film *Apollo 13*.

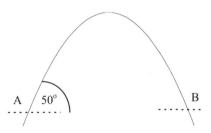

Figure 78. A free-fall trajectory

1) Explain what is meant by a free-fall trajectory and show that in general it takes the form of a parabola.

2) In figure 1 the aircraft enters a free-fall trajectory at A with velocity $v_0=130$ ms^{-1} and leaves it at B. What is the duration of the weightless phase and what is the height of the peak of the trajectory above A?

3) For a given entry velocity what angle at A would maximise the time of free-fall flight.

4) Suggest why the free fall trajectory in the figure is used even though it does not maximise the free-fall time.

3   Power in space

A way of generating power for a spacecraft in an equatorial Earth orbit is to deploy a metal sphere away from the spacecraft on the end of a long conducting tether. As the satellite orbits the tether cuts the Earth's magnetic field lines and a

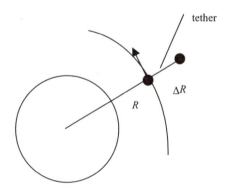

Figure 79. Satellite in Earth orbit of radius R with a tether of length ΔR

voltage is developed across it. A recent trial of the method was conducted from the Space Shuttle *Columbia* in February 1996 using a tether 20 km long at an altitude of 230 km.

1) Given that the Earth's magnetic field is 5×10^{-5} W m^{-2}, estimate the voltage such a length of tether would generate.

The voltage can drive a current by attracting electrons from the ionosphere on to the sphere at one end of the tether whilst electrons are ejected from the space shuttle at the other end by means of an electron gun.

2) Calculate the drag force operating on the tether when a current of 0.5 A flows through it.

3) With this current how much of the power is available for use in the Shuttle given that the resistance of the cable is $0.12 \ \Omega \ m^{-1}$?

4) Where does the energy obtained come from?

5) Estimate for how long this amount of power could be extracted. (The mass of the Space Shuttle is 105 tonnes.)

6) If a current were driven through the tether in the opposite sense what effect would this have on the Shuttle and its tethered satellite?

7) The mass of the tethered sphere is 518 kg, the mass per unit length of the tether is 8.2×10^{-3} kg m^{-1} and the breaking strength of the tether is 1780 N. What margin of strength does this give when the tether is fully extended? (In the 1996 test the tether broke because of an electrical fault.)

8) What effect does the extraction of power have on the angular momentum of the combined shuttle and tethered satellite?

4   Tethered satellites

In this question we look at some more advantages of the tethering of orbiting bodies (see question 3). The centre of mass of the combined Shuttle and tethered satellite lies between them so the satellite goes too fast for a Kepler orbit at its radial distance and the Shuttle too slowly. This gives the Shuttle a small amount of 'gravity'.

1) In which direction is the 'floor' of the Shuttle? Estimate how long it would take an object to fall to the floor. [Hint: look at the system from a rotating frame.]

2) Suppose that after a rendezvous with Mir (or the International Space Station) the Space Shuttle is lowered on the end of a tether towards the top of the Earth's atmosphere where it uncouples prior to re-entry. How would this manoeuvre affect the orbit of the Space Station and what advantages would it bring?

5   The fate of the Earth

When the Sun ends its main sequence life it will swell in size to become a red giant. If the Sun suffers no appreciable mass loss in this transition then it will engulf the Earth (*Rev. Mod. Phys.*, 1997, **69**, 341).

1) Estimate roughly the decay time of the Earth's orbit in the photosphere of a red giant where the density is 10^{-3} kg m^{-3}.

2) However, recent work suggests that the swelling Sun could lose as much as 30% of its mass in the form of a strong solar wind. What change in the Earth's orbit would such a mass loss produce? (This may be sufficient to avoid the Earth being engulfed.)

6   Perfect rockets

In a conventional rocket using chemical fuel the velocity of the exhaust gas is constant relative to the rocket. Consider instead a 'perfect rocket' in which the exhaust velocity u at any time is equal to the the the initial exhaust velocity u_0 plus the current velocity v of the rocket; i.e. $u = u_0 + v$. Thus viewed from the initial rest frame of the rocket the exhaust gases have the same velocity throughout the flight (Fig 80).

1) Show that the efficiency of a perfect rocket, of initial total mass m_0 and final total mass m is given by $(1 - m/m_0)$, where the efficiency is defined as (final K.E. of the rocket)/(final K.E. of rocket + exhaust gases).

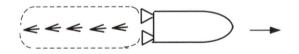

Figure 80. The rocket and its exhaust gases in the initial rest frame of the rocket

1) Show that if the power of a perfect rocket is constant then its acceleration is constant.

2) Explain why a perfect rocket is more efficient than the usual chemical rocket. (Hint: Draw the figure for the chemical rocket corresponding to figure 80 and compare the two figures.)

7   The speed limit for wheel-driven vehicles

The highest speed achieved by a vehicle driven through its wheels at the present time (June 1998) is 439 mph (702 kph) averaged over one kilometer. The car that achieved this feat, the *Turbinator*, was powered by a gas turbine engine which had a maximum power output of 2.8 MW (3750 hp). The car had a frontal area of about $1m^2$, a drag coefficient of about 0.2 and a mass of 1534 kg. The drag on a body of cross sectional area A moving with speed v through a medium of density ρ is $\frac{1}{2}C_D\rho Av^2$ where C_D is the drag coefficient.

1) Estimate the theoretical top speed of the *Turbinator* on an ideal track of unlimited length.

2) Explain why increasing the maximum power output of the engine of a car driven through its wheels, beyond a certain point, cannot lead to any increase in its top speed.

3) What is the value of this limiting power and what theoretical top speed would it give?

4) The speeds calculated above are both much larger than the speed actually reached by the *Turbinator*. The Bonneville salt flats, where the speed run was made, provide a usable track length of about 11 km. Estimate the theoretical top speed that can be reached by a wheel-driven vehicle on this track.

8   The universality of free-fall

To test the principle that all bodies fall with equal acceleration under gravity (the universality of free-fall) Galileo is supposed to have dropped weights from the tower of Piza. In order to test whether this is a realistic experiment a lead musket ball (mass 50g) and a lead cannon ball (mass 10 kg) are released simultaneously from the top of the 50 m high tower of the Leicester University Engineering Building, a well-known landmark in central England. The under-surfaces of the balls are equidistant from the ground at release.

1) Explain why they will not hit the ground simultaneously. Which one will hit the ground first?

2) Show that both balls will be falling with approximately constant acceleration up to the point where they hit the ground (i.e. phase 1 of question 30).

3) From an approximate solution of the equation of motion estimate the time interval separating their impacts with the ground. Would the departure from free-fall have been detectable with seventeenth century technology?

9   Bubbles

1) Why do the bubbles in a glass of champagne travel upwards? Explain where the energy comes from.

2) A coupe of champagne is placed at the centre of a spinning turntable. Draw a figure to show the directions in which the champagne bubbles will move.

10   The deep blue sea

The following question was asked by a reader of the *Independent* newspaper: 'How long would it take something to fall to the bottom of the deepest ocean?' The deepest part of the sea is the Marianas trench where the floor is 10.9 km below sea level.

1) For an object that is appreciably denser than water show that the time of fall is inversely proportional to the square root of the linear size of the object.

2) Estimate the time of fall for a stone that can be held in the hand.

3) Show that an incompressible object with a density less than 1080 kg m^{-3} would not come to rest on the bottom.

The density of sea water is 1025 kg m^{-3}.
The bulk modulus of water is 2.1×10^9 N m^{-2}.

11   All hail

1) Large hailstones can destroy grapes and even the vines themselves. Show that the damage a hail stone can inflict increases as the fourth power of its linear size and estimate the constant of proportionality.

2) In Texas severe hail storms occur in which individual hail stones reach a diameter of 10 cm. It is said that the kinetic energy of such a hail stone exceeds that of a baseball travelling at 80 mph. Estimate the ratio of the two kinetic energies.

The mass of a baseball is 0.168 kg.

12   Escape into space

The concept of a perfect rocket (question 6) could not be applied to lifting rockets from the surface of the Earth into space since this requires the high power which is available from chemical fuels. However the ability of the conventional rocket to reach escape velocity from the surface of a planet can be improved by the following technique. Instead of launching the rocket vertically, it is launched horizontally along a low friction track. Once it has acquired a sizeable horizontal speed the rocket is steered to the vertical by curving the track.

1) Show that in the absence of air resistance a higher terminal velocity can be achieved by this technique than for a vertical launch from a given mass of fuel.

2) Explain why this method of launching a rocket is more efficient than the usual way.

3) Why would the technique offer less of an advantage in the presence of air resistance?

13   In orbit

Consider the following problem.

> If in response to Joshua's command the Earth had really been stopped in its orbit how long would it have taken for the Earth to fall into the Sun?

It is often a good idea to look for a short cut to solve a problem which would otherwise involve heavy algebra. The following cut is however too short.

> By stopping the Earth only the tangential motion is affected, the radial motion being unchanged. The Earth will therefore fall into the Sun in the time it takes to complete one quarter of an orbit.

1) What is wrong with this reasoning?

2) This problem can be solved correctly by a short cut using Kepler's third law. This is the law that relates the orbital period T about a body of mass M to the semi-major axis of the orbit, a, $T = 2\pi a^{3/2}/\sqrt{(GM)}$. This formula does not involve the minor axis, so can be used for any orbit, including the radial one in the limit $b \rightarrow 0$!
Construct an argument to show that the Earth falls into the Sun in $3/\sqrt{2}$ months.

14   Journey to Mercury

There has been only one landing of a spacecraft on the planet Mercury, by Mariner 10 in 1974/5. Show that a one way trip to Mercury requires more fuel than is needed for a journey to any other planet in the Solar System. The radius of Mercury's orbits is about 0.39 AU.

15 Not so square

From time to time it has been suggested that gravity on a sufficiently large scale might not follow an exact inverse square law. For example, in the nineteenth century this was proposed to explain the anomalous perihelion advance of Mercury (N.T. Roseveare, *Mercury's Perihelion from Le Verrier to Einstein*, Oxford, 1982) and more recently to account for the dynamics of clusters of galaxies without invoking dark matter (Milgrom, M., *Ann. Phys.*, 1994, **229**, 384).

1) Which of Kepler's laws would remain valid if the inverse square law were replaced by a different central force law?
   The inverse square law is a rather special force law in several regards. To see this we can investigate the inverse cube law. You may assume the orbit equation for a central force F in the form
   $$\frac{d^2u}{d\theta^2} + u = -\frac{F}{mL^2u^2}$$
   for a particle mass m, angular momentum per unit mass L at the point with polar coordinates $(1/u, \theta)$. For an inverse cube law let $F = -ku^3$, where k is a constant.

2) Show that $k = mL^2$ leads to either a circular orbit or spiral orbits.

3) If a body in a circular orbit under an inverse cube law is perturbed what will happen to it?

4) Describe the other possible orbits for $k > mL^2$ and $k < mL^2$.

5) Show that there are no closed stable orbits with an inverse cube law.

6) A gravitating spherical body acts like a point mass at its centre as far as the force it exerts on an external body is concerned.

(a) Would this result be true if gravity were described instead by an inverse cube law?

(b) Would it be possible to describe the action of an extended spherical body by a point source in this case?

16 The tractor beam of the *Enterprise*

The *Starship Enterprise* is equipped with a tractor beam that can lock on to a drifting spaceship. Once the beam has locked on, the spaceship follows a *circular* path which ends in the cargo bay of the *Enterprise.* The angular momentum of the tractored spaceship, with respect to the *Enterprise,* is constant. How does the tractor force fall off with distance?(A. P. French, *Newtonian Mechanics*).

17   Orbits inside matter

Newton, in a letter to Hooke in 1679, proposed an experiment to demonstrate the rotation of the Earth. He reasoned that a body dropped from the top of a tower would fall to the east as the tangential velocity of the top of the tower, due to the Earth's rotation, is greater than that of the ground. Newton claimed that this deflection could be measured and included a sketch purporting to show the path of the body above the ground and also the trajectory it would follow if it were to continue to fall without resistance into the Earth. The path he drew was a spiral (fig 81) ending at the Earth's centre This spiral orbit was a rare mistake

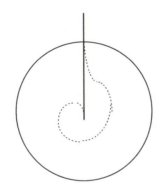

Figure 81. Newtons erroneous spiral

on Newton's part and he was deeply chagrined when Hooke corrected him (R Westfall, *Never at Rest*).

1) Determine the orbit that a body released from rest at the equator would follow if it could fall without resistance into the Earth, assuming that the density of the Earth is constant.

2) What is the distance of closest approach of the body to the centre of the Earth?

3) Possible candidates for a body that could penetrate the Earth are WIMPs (Weakly Interacting Massive Particles) and black holes. Black holes of sufficiently low mass M have a geometrical interaction cross-section πR^2, where $R = 2GM/c^2$. By comparing the magnitude of the force between a black hole and a nucleus with the interatomic force show that this cross-section is

appropriate for a black hole of mass 10^3 kg.

4) Estimate the mean free path of such a hole.

5) However, black holes emit blackbody radiation (Hawking radiation) at a temperature $T = hc^3/(8\pi GkM)$ from a surface of area $4\pi R^2$. Show that a hole of mass M has a lifetime

$$t = 1.7 \times 10^9 \left(\frac{M}{10^{12} \text{ kg}} \right)^3 \text{ years}$$

and hence that a black hole of mass 10^3 kg would not survive long enough to complete a single orbit.

18   The Indian rope trick

In the Indian rope trick a rope is thrown into the air; the rope remains magically suspended while someone climbs up it and disappears into the sky. The rope is then retrieved. This trick is ususally reported by observers to be performed late in the evening. The purpose of this assignment is to show that it is in fact possible in theory to position a rope in apparent suspension in this way - although it needs a very long, strong rope and not in India!

1) A satellite placed in a circular orbit above the Earth's equator moves synchronously with the rotation of the Earth. What is the radius of the satellite's orbit?

2) A long thin rigid cylinder is placed in orbit about the Earth with its long axis pointing towards the Earth's centre. Give an argument to show that its attitude is stable.

3) A satellite consists of a long rope of constant mass per unit length placed in orbit above the equator which hovers over a fixed point just above the ground and stretches radially outwards into space. Calculate the distance of the far end of the rope from the Earth's centre. (Hint: view the system from a rotating frame in which both Earth and rope are stationary. Equate the total inward force on the rope due to gravity to the total outward centrifugal force.)

4) At what position along the rope will its tension be a maximum?

5) What is the magnitude of this tension?

6) What must the Young's modulus of the rope be if it is not to break?

7) Discuss quantitatively the feasibility of such a satellite. Young's modulus for strong steel, for example, is 2×10^{11} N m^{-2}.

The rope can be used to support a space elevator which would lift payloads into geosynchronous orbit. This idea was first proposed by the Russian engineer Y. N. Artsutanov in 1960 (see A. C. Clarke's novel *The Fountains of Paradise*).

8) Calculate the cost of electricity used in transporting a kilogram into geosynchronous Earth orbit by this method. The cost of electricity in the UK is currently about £0.08 per kW hour.

19 Time and tides

The Moon raises tidal bulges in the Earth's oceans which, because of friction between the rotating Earth and the water, are displaced with respect to the line joining the centres of the Earth and Moon in the direction of the Earth's rotation. The Moon opposes this displacement and hence, exerts a braking torque on the rotating Earth. We ignore the effect of the Sun here. The following additional

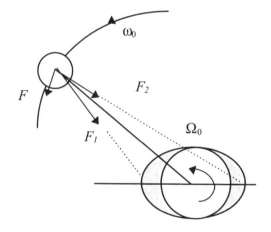

Figure 82. The displaced tidal bulges exert forces F_1 and F_2 on the moon. The resultant force F accelerates the moon in its orbit. In the figure the Earth is viewed from the north pole.

data is required: Moment of Inertia of the Earth $= 8.07 \times 10^{37}$ kg m^{-2}

Mass of the Moon $= 7.35 \times 10^{22}$ kg

Moon's orbital radius $= 3.85 \times 10^8$ m

Orbital angular velocity of the Moon $w_0 = 2.64 \times 10^{-6}$ s^{-1}

Angular velocity of the Earth about its axis $= 27.5\, w_0$

1) Is the angular momentum of the Earth-Moon system conserved?

2) Explain why the Moon moves away from the Earth?

3) Eventually the process will come to an end. Why?

4) What will be the period of the Earth-Moon system at this time?

5) From lunar ranging measurements the present rate of retreat of the moon from the Earth is 3.82 cm yr^{-1}

(a) At this rate about how long will it take for the process to come to an end?

(b) What determines the timescale of this process?

6) Suppose that the Earth were turning on its axis more slowly than the Moon but in the same sense. What would the action of tidal friction be on this system?

7) Suppose the Earth were rotating in the opposite sense: Describe the action of tidal friction on the system. What would be the endpoint of the evolution of such a system?

8) Now consider the action of the Sun on the Earth tides after tidal lock between the Earth and the Moon has been achieved. What will be the fate of the Earth-Moon system?

20   The day the Earth caught fire

Edward Teller has been an advocate of a programme to develop, from star wars technology, the capability to deflect a large meteorite which might happen to be on a collision course with the Earth. The meteorite would be intercepted by one or more rockets carrying nuclear warheads which would be exploded to deflect the course of the meteorite by enough for it to miss the Earth. The figure shows the orbits of the Earth and the meteorite on a collision course. The Earth and meteorite are shown at the instant when the warheads explode.

1) Investigate the feasibility of Teller's scheme by estimating the deflection that could be achieved with a large nuclear weapon using the following data.
10^6 tons of TNT (1 Mt) $= 4.2 \times 10^{15}$ J.
Yields of powerful nuclear weapons are in the range 1-100 Mt.
The binding energy of meteoritic rock is of order 1 eV per atom.

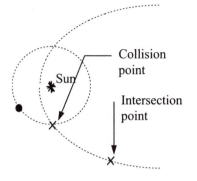

Figure 83. An asteriod approaches the Earth

A 12 km diameter meteorite may have caused the extinction of the Dinosaurs (Physics Today, July 24 1987) at the end of the Cretaceous epoch. It would have made a crater of the order of 100 km in diameter, where the crater size is proportional to (energy of meteorite)$^{1/3}$. The relationship between the size of the meteorite and the mean time between collisions of the Earth with such an object (the waiting time) is waiting time $= 10^6 \times$ (diameter in km)^2.

2) Would nudging the meteorite in the direction of its orbit be any more efficient than exerting a transverse force?

21 Space debris

In addition to about 550 operational spacecraft in Earth orbit there are an estimated 9500 other pieces of junk from previous space missions which are large enough to be tracked from the ground. A collision with any of these objects, which are about 10 cm and larger in size, would destroy a spacecraft. The number density of this debris peaks at a value of about 10^{-17} m^{-3}. Estimate the probability that a spacecraft of cross sectional area 30 m^2 will be hit in its operational lifetime of 5 years if it is located in the region of peak density. (*Scientific American*, August 1998).

22 Disposing of space junk

From the London *Times* (Mon Oct 28 1996)

[To dispose of pieces of space junk between one and ten centimetres long,] a ground based laser would ...using short sharp pulses, burn off a portion of the underside. The evaporating stream of material would act as a thruster, nudging the particle from its circular orbit round the Earth into a more elliptical one. The particle's new orbit would take it into [the outer fringes of the] atmosphere where it would [eventually] burn up safely.

The object of this problem is to calculate the laser power required assuming adaptive optics can be used to limit the spread of the beam to a circle 1m in diameter.

1) Consider a piece of debris in a circular orbit at 400 km altitude. Show that the new orbit which results from imparting an outward radial component of velocity to the junk has a perigee at an altitude of less than 400 km. Hint: Is there any change in the angular momentum of the junk?

2) Show that adding a radial component of 300 km s^{-1} is enough to take the junk below 150 km where significant atmospheric drag will occur.

3) Estimate the minimum energy that has to be supplied by the laser to a 100 gm piece of aluminium. The heat of vapourisation of aluminium is 10^6 J kg^{-1}.

4) Discuss the feasibility of using a single short pulse to accomplish this.

According to the *Times* report this is within current technology and would cost $50 -100 million. It would make the Space Station safe from impacts.

23 Suspension bridges

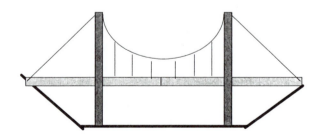

Figure 84. A suspension bridge

Figure 84 shows a suspension bridge. The roadway is hung from two multi-stranded high tensile steel cables which pass over towers to anchor points in the ground at each end. The Akashi Kaikyo road bridge (opened in 1998) has a span of 1990 m and the cables each have a diameter of 1.12 m. In this question we investigate how much scope there is for building suspension bridges with even longer spans.

1) At which point is the tension in the cables of a suspension bridge greatest?

2) Estimate the maximum length of a bridge which has a roadway mass of 30 tonnes m^{-1} and a cable diameter of 2 m. The breaking stress of high tensile steel is 2×10^9 Pa.

3) Estimate the upper limit to the length of such a bridge by taking the mass of the roadway to be negligible compared with the cable mass.

24   Tennis then and now

In the 1970s it was believed that, for a given speed of impact, the tauter the strings of a tennis raquet the more powerful the shot. This is now known to be false: if the strings are tightened beyond a certain tension the speed of the ball off the racket is reduced. Provide an explanation for this fact.

25   Energy in life

Energy is released within the body by oxidation: 1 ml of O_2 releases 20 J. For most of the time our intake of oxygen is in balance with the rate at which our bodies generate energy: this is called aerobic energy production. For short periods when high muscular effort is required the body can operate anaerobically with the rate of energy production exceeding the oxygen intake. This creates an oxygen debt which is repaid by our being out of breath for a while after the effort. Only about 25% of the energy released during the activity can appear as mechanical work. (The rest appears as heat.) Even when at rest the body must expend energy to remain alive. This energy production appears as heat and is called the basal metabolic rate (BMR). Its value for a body of mass M kg is $4.06M^{0.68}$ W.

1) What oxygen intake in ml kg^{-1} min^{-1} is required to support your BMR?

2) Given that it is possible to survive for a brief time only at the summit of Mt. Everest at a height of 8.5 km, estimate the length of time you could survive on the oxygen content of a typical household room.

3) The maximum continuous oxygen intake achievable by a trained athlete is about 80 ml kg^{-1} min^{-1}. Activities like competitive rowing and human powered flight require a high power output over a lengthy period. What is the maximum rate of working in horsepower (= 755 W) that a 60kg human can sustain during such an activity?

4) A sprinter needs the energy from approximately 10 l of O_2 to run 100 m in 10 s.

(a) Estimate the useful power output in horsepower of the sprinter during the race.

(b) Is sprinting an aerobic or an anaerobic activity?

5) On a television programme two men claimed that they could lift a 5.5 ton armoured car 1 m off the ground in less than one minute using their own muscular effort transmitted through a system of pulleys. The audience were asked to predict the outcome of the attempt. In fact, on this occasion the men failed. Decide whether such a feat is humanly possible.

26   Hot potatoes

In the everyday world a hot object, for example a hot potato, placed in a uniform environment will lose heat at a rate governed by Newton's law of cooling, according to which the rate of loss of heat is proportional to the temperature difference ΔT between the surface of the potato and its surroundings. (This is an empirical law but quite a good one here.) The flow of heat within a hot body is governed by the diffusion equation

$$\frac{\partial \Delta T}{\partial t} = D\nabla^2 \Delta T$$

where D is the diffusion constant.

1) Indicate what the temperature distribution inside a hot potato will look like.

2) Show that $d\Delta T/dt = -\Delta T/\tau$ where τ is the thermal time constant. Show that τ is the time for the temperature difference to drop by a factor e. (Note that if the surroundings are hotter then this same time constant applies to the heating process.)

3) Suppose that you have invited guests to dinner and wish to serve them white wine chilled to the recommended temperature of 8 °C. If the room temperature is 22 °C and the temperature of your fridge is 5 °C how long before the arrival of the guests should you put the wine into the fridge? (Take the time constant of a bottle of wine to be 1 hr.)

4) Returning to the potato, use the diffusion equation and the result of question (2) to show that $\tau \propto M^{2/3}$, where M is the mass of the potato. [Hint: estimate $\nabla^2 \Delta T$ in terms of the size of the potato.]

5) Cookery books recommend that meat be roasted for 30 mins plus 30 mins per pound. Assuming that meat cooks as soon as it has reached a certain

temperature, how do these instruction accord with the result of the previous question?

27 Thermal balance in mammals

The rate at which an animal looses heat to its surroundings is also governed by Newton's law of cooling (see question 26). The rate at which an animal generates heat is governed by its activity, but for a mammal cannot fall below its BMR. Most mammals must maintain their core (interior) body temperature at a steady value in the region of 37 C. They do this by balancing heat generation against heat losses. A variety of strategies are employed, for example sweating to increase heat loss.

1) Suggest other mechanisms for maintaining body temperature.

2) Imagine that you are enjoying a relaxing day on holiday and the ambient temperature is 37 C in the shade. Using the following data estimate the minimum quantity of water that you must drink over a 10 hour period to maintain your body temperature at its normal value. (Latent heat of evaporation of water is 2.4×10^6 J kg^{-1}.) (Metabolic rates for a body of mass M in units of $M^{0.68}$ W are: sitting: 6.01; standing 7.34; walking 24.7.) Why is your estimate a minimum?

3) Small animals are at a disadvantage in a very cold climate and also in a very hot one. Explain why this is.

4) Camels do not sweat. Suggest how they cope with the extremely high daytime temperatures of the desert. (Note that the desert is cool at night.) Check your answer numerically.

28 Over(h)eating

Health farms offer the following treatment to slimmers. The slimmer is coated with wax and then swathed in layers of sheeting, the purpose being to prevent heat loss and so induce copious sweating.

1) Show that the human body temperature increases by about 1°C per hour if it cannot lose heat.

2) How long could a person survive the absence of any heat loss at all? (The BMR for a body of mass M kg is $4.06M^{0.68}$ W; see question 25.)

3) How effective do you think the treatment is as a slimming cure?

As actually carried out in practice the slimmer's head is not covered, so some heat loss can occur.

4) Estimate the weight loss that can be achieved in a week through fasting assuming that the energy content of fat is 24 MJ kg^{-1}.

29 Dinosaur metabolism

1) Palaeontologists are interested in whether dinosaurs were warm or cold blooded (ie did they have a mammalian or reptilian metabolism?) Mammals and birds have regulating mechanisms which maintain their core body temperatures at a steady value day and night so they are classified as endotherms. Modern reptiles do not have this ability to regulate their body temperature; they are ectotherms. When resting their temperature depends on their surroundings so in cool conditions their body temperature tends to fall (and vice-versa). Vital chemical processes slow down as body temperature falls, so in cool conditions ectotherms become sluggish - they move and digest their food slowly. Although they lack internal regulation ectotherms are able to control their body temperature to some extent by their behaviour e.g. in daylight hours by moving back and forth between sun and shade. R. McN Alexander has estimated the time constant of a 50 tonne dinosaur to be 20 days. The largest sauropod dinosaurs probably reached a mass of 100 tonnes (see Science, **266**, Dec 1994). Estimate the thermal time constant of such an animal. (See question 26.)

2) Show that such a dinosaur, if it possessed a mammalian metabolism, would be in danger of cooking itself through from inside to out. (Note that the Earth's climate in the mesosoic era was tropical/subtropical).

30 Dropping like lead

The lead shot used in shotgun cartridges consists of small spherical pellets 2-3 mm in diameter. The pellets are made by pouring molten lead through a frame pierced with holes which is suspended over a long drop inside a shot tower. The falling lead forms into small spherical drops which solidify and are collected in a tank of water at the bottom of the tower. The size of the shot produced is determined by the size of the holes in the frame and the size of the shot determines the drop height that must be used. Accurately spherical shot has been produced by this method since its invention by William Watts in 1782. In the following you will need to know that the drag force on a falling object is $\frac{1}{2}C_d\rho Av^2$ where v is the velocity of the object, ρ is the density of air, A is the cross-sectional area of the body and C_d is a constant of order 0.5 for a sphere.

1) Show that the fall may be divided approximately into two stages: stage 1 a phase of constant acceleration and stage 2 a phase of constant velocity.

2) Find the duration of stage 1 for a pellet of mass m and cross-section area A.

3) Explain why the molten shot must solidify before the end of stage 1 if spherical shot is to be produced. [Hint: How does the *weight* of the falling shot vary?]

4) Assuming that a molten lead droplet looses heat through Newton's law of cooling, explain why there is an upper limit to the size of lead shot that can be produced by this dropping method. In this context Newton's law of cooling states that the rate of loss of heat is proportional to the surface area of the droplet × the temperature difference between the drop and its surroundings. For the forced cooling in air the constant of proportionality is about 200 W m^{-2}K^{-1}.

5) Estimate this upper limit. [A typical shot size, number 5 shot, has a mass of 1.27×10^{-4} kg. Latent heat of solidification of lead $= 2.31 \times 10^4$ J kg^{-1}. Melting point of lead $= 327.5$ C. Density of lead $= 11\,300$ kg m^{-3}.]

6) Someone has the 'bright idea' of removing the air from a shot tower to lengthen the free-fall time. Show that such a tower could not in practice produce spherical shot.

31 Journey to the centre of the Earth

In Jules Verne's novel *Journey to the Centre of the Earth* a party of explorers descend into the crater of an extinct volcano and after traversing many passages and shafts eventually reach the centre of the Earth. Throughout the journey the expedition appears to experience normal gravity.

1) What changes in the force of gravity would such explorers actually have experienced as they descended?

2) Suppose we take Jules Verne's word that the air at the Earth's centre is connected to the air at its surface and is at roughly the same temperature, and assume that the air is an ideal gas. Obtain a value for the pressure at the Earth's centre under these assumptions.

3) The value of the pressure according to the assumptions of part (b) is not believable. Decide what would actually happen if a hole were to be bored to the
   centre of the Earth and atmospheric air allowed to fall in.

32   Buoyancy

1) In one of Arthur C. Clarke's *Rama* novels the occupants of a spaceship are
   subject to a long episode of high acceleration. In order to make this tolerable
   for the human occupants they are given breathing apparatus and completely
   immersed in a fluid, in which they have zero buoyancy. How effective do you
   think this strategy would be?

2) All objects exhibit diamagnetism, so given a sufficiently large field gradient
   any object, including living creatures, can be levitated. This has been demonstrated by levitating a mouse near the top of the solenoid of a Bitter magnet.
   In principle human beings could be levitated in this way given a sufficiently
   large volume and strength of magnetic field. Would a person undergoing such
   levitation experience true weightlessness? (Physics Today, Sept 1998, **51**.)

33   Survival on Venus

Suppose a feasibility study is being conducted into a spaceship to land people
on the surface of Venus for a brief time, where the surface temperature is 475 °C
and the surface pressure is 95 Earth atmospheres.

1) You are given the task of looking into the cooling requirements of the crew's
   quarters which are a cylinder of internal diameter 3m and height 5m. This
   volume is insulated from the Venusian atmosphere by walls 1m thick having
   a thermal conductivity of 0.01 W m^{-1}K^{-1}.

2) Estimate how long after landing it would take for the interior of the spacecraft
   to become intolerable to its occupants if no steps are taken to cool it?

3) How much power would an ideal refrigerator require to maintain a cabin temperature of 20 °C?

4) What is the density of the Venusian atmosphere at the planet's surface in terms
   of the sea-level density of the Earth's atmosphere? [The Venusian atmosphere
   consists mainly of CO_2.]

5) The spaceship parachutes on to the surface with a parachute that would give
   a terminal velocity in the Earth's atmosphere on landing of 10 m s^{-1}. What
   would be the corresponding terminal velocity on Venus?

34   The Gulf Stream

The Gulf Stream has a strong influence on the winter climate of north western Europe because of the very large heat capacity of water. Show that a 2 metre thick layer of the Earth's oceans has a heat capacity that is comparable to that of the whole atmosphere. The specific heat of water is 4.22×10^3 J kg^{-1}K^{-1}.

35   The Antarctic ice cap

A hole bored through 2164 m of ice at Byrd Station in the Antarctic found liquid water at the base of the hole. Given that the rate at which the Earth loses heat from its surface is about 0.06 W m^{-2} and that the mean surface temperature of Antarctica is -50 C show that the presence of liquid water is to be expected under such a depth of ice. The thermal conductivity of ice is 2 W m^{-1} K^{-1}.

36 C_d's

1) In elementary physics courses you are told that the resistance experienced by an oil drop of radius a falling with velocity v through air of viscosity η is given by Stokes's law: $F = 6\pi a \eta v$. On the other hand, you may know that car makers always quote the air resistance in terms of a drag coefficient C_d of magnitude ~0.5. This gives the resisting force as $F = \frac{1}{2}C_d \rho A v^2$ for a body of area A moving through air of density ρ. These two laws are not the same (as is clear from the different dependence on v).
   (a) Estimate the magnitude of both forces for an oil drop and a car.
   The coefficient of viscosity for air is $\eta = 1.8 \times 10^5$ N s m^{-2}.
   A typical oil drop has a radius 2 μm.
   (b) Hence explain why one law is used in one case and the other in the other.
   (c) The C_d law is unconnected with viscosity. What is the origin of the drag in this case?

The ratio of the two forces is (a numerical coefficient) $\times \rho a v / \eta$. The ratio $R = \rho a v / \eta$. is called the Reynolds number. For large Reynolds number flows viscosity can be neglected and for small Reynold's numbers inertial drag may be neglected. Reynolds showed that different sized objects of the same shape have the same C_d provided that the Reynolds number is the same.

1) It is desired to measure the drag force on a golf ball at speeds up to 150 mph in a wind tunnel which operates with a maximum air speed of 40 mph. What must the size of the model golf ball be? Note that it is cheaper to work out the answer to this question than to build a bigger wind tunnel.

2) C_d can be assumed not to vary with R over a large range ($10^3 < R < 10^5$ for a smooth sphere). It is claimed that a mouse can survive a fall from any height. Given that a sky diver falling without a parachute reaches a terminal velocity of 120 mph decide whether this claim can be correct?

3) It has been pointed out (Nature, Correspondence, 1994) that the world one hour cycling record, which stands at 53.040 km, could be improved by con-ducting the attempt on an indoor track in a depressurized stadium filled with pure oxygen at a pressure of 0.2 atmospheres. The effect of this environment would be to reduce the air drag acting on the cyclist but should not affect their power output.
(a) Show that the speed of the cyclist is proportional to (density of air)$^{-1/3}$
(b) Hence, what would the world record be on such a track?

4) Prior to 1838 it was believed that a ship powered by steam alone could not cross the Atlantic as it would be unable to carry enough coal in its bunkers for such a long voyage. According to one particular expert, Dr Dionysius Lard-ner, a writer of popular science books, there was a maximum range because the resistance to motion increased in proportion to the coal carried.
(a) Explain the argument and point out the fallacy in it.
(b) Show that, ceteris paribus, the range of a steam ship is proportional to the ship's linear size.

37   Fluid flows

1) In Zola's *Thérèse Raquin* the two lovers murder the husband in a boat on a lake. What happens to the surface level of the lake when they dump the body overboard (assuming that it sinks)?

2) A canal crosses a valley on an aqueduct. Beyond the aqueduct a barge is launched into the canal and is then sailed across the aqueduct. Sketch (or describe) the additional load on the supporting pillars of the aqueduct caused by these operations as a function of time.

3) Estimate roughly the speed of the wind require to lift Dorothy (and her house) on its way to the Land of Oz. The density of air is 1.2 kgm^{-3}.

4) (a) Why must the speed of the blades of a windmill be less than the wind speed (in constrast to the speed of a sailing boat - see the following question)?
(b) What is the maximum power that can be developed by a windmill of effective blade area 5 m^2 in a wind of 10 ms^{-1} ?

5) At the surface of Venus the pressure is 95 Earth atmospheres and the temper-ature is 475C. Ignoring the practical difficulties, is sustained person-powered

flight possible in the Venusian atmosphere, given that, because of the power requirements, it is only marginally possible on Earth?

38   Land yachts

The sail of a yacht is like the wing of an aircraft except that it is oriented vertically rather than horizontally, so an air flow across it produces thrust instead of lift. A yacht uses wind as the source of its motive power. As it gains speed its motion through the air increases the airflow across its sail and it is able to reach speeds in excess of the wind speed. Land yachts which are just yachts on wheels can reach speeds of 3 to 4 times the wind speed.

1) Draw a diagram of the forces acting on a land yacht and explain how its top speed is determined.

2) If θ is the angle between the direction of motion and the velocity of the wind relative to the yacht show that at maximum speed $\sin \theta \propto a/A \ll 1$, where a is the area of the yacht facing into the wind and A is the area of the sail.

3) If the speed of the wind relative to the ground is w show that the maximum speed of the yacht is of order $w \cos^2 \theta / \sin \theta$.

4) At present (August 1998) the world record for a land yacht stands at 151 km hr^{-1}. However it is hoped that a new design, called *Windjet,* which uses a rigid sail and aerofoils to produce negative lift (a technology used in formula 1 motor racing) will raise the record beyond 160 km hr^{-1} (100 mph). Explain why increasing the effective weight of a land yacht could increase its top speed.

5) The team behind the new land yacht believe that on ice their yacht could achieve a speed of 240 km hr^{-1}. Explain the thinking behind this belief.

39   Bouncing bombs

In the second world war a new type of bomb was invented by Barnes Wallis for use against dams. The bomb was launched from a low-flying aircraft on to the surface of the dammed water in the direction of the retaining wall. For a sufficiently shallow angle of incidence the bomb bounced along the surface of the water, like a skimming stone, until it struck the dam wall. The bomb then sank to the base of the wall and was exploded by a depth sensitive detonator. Navel gunners of the 17th and 18th centuries had used this effect to increase the range of their guns. Projectiles will bounce off water only if the angle of incidence with the water is smaller than a certain value. Experiments carried out by Ramsauer in 1903 showed that non-rotating brass spheres fired at a water surface ricochet off provided that the angle of incidence, measured to the water surface, is less than 7° (*Handbook of Ballistics*, C Cranz and K Becker). Below 7° the smaller the angle of incidence the larger the number of bounces that follow.

1) Show that the maximum angle of incidence for which bouncing can occur is inversely proportional to the square root of the density of the sphere. Hint: The pressure exerted on an incident sphere with speed v by water of density ρ is proportional to ρv^2.

2) Barnes Wallis's bomb had a density of about 2500 kg m^{-3} and was delivered from a plane flying at 111 m s^{-1} at a height above the water of 18m. Show that this method of delivery ensures the bomb will bounce on the water. The density of brass is 8400 kg m^{-3}.

3) Barnes Wallis found that imparting a backspin to the bomb before it was dropped conferred several advantages. Explain how, other things being unchanged, backspin lowers the angle of incidence of the bomb with the water and increases the range of the bomb both while it is dropping and while it is bouncing.

4) Another reason for backspin was that it caused the bomb to be deflected downwards by the collision with the dam wall rather than climbing up the wall and going over the top. Show that a spherical ball of radius a with backspin ω that strikes a vertical wall horizontally will acquire a downward component of velocity $2a\omega/7$.

5) It was necessary for the bomb to be in contact with the base of the dam wall when it exploded. Explain how this was achieved despite the fact that the bomb rebounded from the wall after striking it. Note that the sinking bomb still possessed some of its backspin.

40   Mirrors without the grind

Traditionally a telescope mirror is made by casting a flat-faced blank of glass and then laboriously grinding away material from the face to shape it into a paraboloid. At the University of Arizona, a telescope mirror 6.5 m in diameter has been cast in a mould rotating at a steady 7.4 rpm.

1) Show that the surface of a liquid rotating with constant angular velocity forms into a paraboloid. Hint: Consider the equilibrium of a particle at the surface of the liquid.

2) What is the f-number for this mirror?

41 Sand and water

1) Show that the mass flux of sand particles of density ρ through an open pipe of radius a from a large reservoir can be obtained by dimensional analysis as
$$\dot{M} \propto \rho g^{1/2} a^{5/2}.$$

2) The flux of incompressible fluid of density ρ through a pipe of radius a is obtained from Poiseuille's formula as
$$\dot{M} \sim \frac{\rho g a^4}{c_s \lambda},$$
where λ is the mean free path of fluid particles and c_s is the sound speed in the fluid. Explain why this cannot be obtained by dimensional analysis.

42 Thunder and lightning

In fine weather there is a vertical electric field of about 100 V m^{-1} above the Earth's surface which connects positive charge in the upper atmosphere to negative charge on the Earth.

1) Estimate the amount of this negative charge on the Earth.

It is believed that thunder storms around the globe are responsible for maintaining this potential difference by transferring negative charge from thunder clouds down to ground.

2) In a typical thunder storm the voltage difference between the clouds and ground is about 200 MV. On average there are say four flashes of lightning per minute each of which transports about 30 C of negative charge to Earth.

3) Estimate the power output of the storm.

Each flash consists of about three strokes which are principally responsible for transferring most of the charge to Earth. A stroke lasts for approximately 50 ms and heats the air along its track to a temperature of about 30,000 C.

4) Estimate the energy dissipated by a stroke.

5) Hence, what is the width of the conducting (i.e. ionised) track down which the discharge takes place? Take the altitude of the region of negative charge in the thunder cloud to be 4 km.

6) What is the source of the energy which powers a thunder storm?

7) Estimate the speed at which the column of air, heated by a lightning stroke, expands into the surrounding air.

8) Compare this expansion speed with the speed of sound in the atmosphere and hence explain the production of thunder.

9) The average acoustic energy generated by a flash of lightning has been estimated to be 3×10^6 J (Holmes et al., 1971, J. Geophys. Res., **76**, 2106). What is the efficiency of conversion of flash energy to acoustic energy?

43   Diverging beams

1) The bunches of particles in an accelerator will expand laterally because of the mutual Coulomb repulsion of particles with like charges as they move between the focussing magnets situated at intervals around the accelerator ring. In the *Tevatron* proton accelerator at Fermilab the particles are injected in bunches of 2×10^{11} in the form of cylinders of radius of 36×10^{-6} m and length 0.5 m. Obtain a relationship between the radius of the bunch and time in its *rest frame* in the absence of focussing.

Note: $\int_1^2 \frac{dx}{(\log_e x)^{1/2}} \simeq 2.145$.

2) How long does it take for the beam to double in radius in its rest frame.

3) Fortunately relativity comes to the rescue of accelerator engineers. Given that the laboratory energy of the protons is 10^{12} eV, over what time and distance in the laboratory frame does the spreading in (2) occur?

44   A poor man's black hole

Imagine a future time when travel to the stars is about to be undertaken. A spaceship capable of sustained acceleration at a rate of $1g$ has been developed (never mind how). The physical well-being of the crew is well catered for by an acceleration of $1g$ but their psychological well being requires that they are able to receive news from home to combat the effects of isolation.

1) By using a spacetime diagram show that after a certain period of time no signal sent from the Earth towards the spaceship will ever be received whereas the Earth can continue to receive messages from the spaceship and that this state of affairs will persist as long as the spaceship continues to accelerate. In effect the Earth has passed through an event horizon and is cut off by the horizon from the spacetime region of the spaceship.

2) If the date of departure of the astronauts is 1 January 2500 after which date will the signals sent fail to reach the spaceship?

3) How long after their departure, measured in spaceship time, will the astronauts loose contact with Earth?

4) Explain how contact with the Earth could be reestablished.

45 The Ur-Star

A visit to other stars in our Galaxy in a human lifetime time would involve very high relativisitic γ - factors. As a result space travellers of the future will, like the three wise men, find themselves following a star. This is because for high enough Doppler factors the cosmic microwave background is aberrated into the forward direction and shifted to visible frequencies.

1) Show that for a traveller whose Lorentz factor γ is 1000 the microwave background is seen with the naked eye as a resolved disc of light which has a blackbody temperature at its centre of 5460K.

2) Estimate the angular size of the visible disc.

For a blackbody the brightness (energy per second per unit area per unit solid angle per unit frequency) is

$$I_\nu = \frac{2h\nu^3}{c^2 \left(e^{h\nu/kT} - 1\right)}$$

and I_ν/ν^3 is a Lorentz invariant. The temperature of the cosmic microwave background is 2.73 K.

46 Hearing things

The speed of sound in a medium of density ρ and pressure P is usually obtained from

$$c^2 = K/\rho \qquad\qquad (73)$$

where $K = -VdP/dV$ is the bulk modulus of the medium, which may be a solid, liquid or a gas and V is an arbitrary volume.

1) Show that the sound speed can also be expressed as

$$c^2 = dP/d\rho \qquad\qquad (74)$$

Hence explain why sound travels faster in water than in air.

2) Use equation (74) to show that in an ideal gas of molecular weight μm_p at temperature T the velocity of sound is proportional to $\sqrt{T/\mu m_p}$ and is therefore independent of pressure and density.

3) Why does the pitch of a person's voice go up when their mouth and lungs are filled with helium gas ? (Do not experiment - it can be fatal.) By how much

is the pitch raised ? Hint: the voice is produced by vibrations in the larynx the dimensions of which fix the wavelength.

4) An ecentric Hi-fi enthusiast listens to a recording of a piece of music in a helium filled room. Would the pitch of the music be affected ?

5) The derranged enthusiast now hires an Orchestra, equips them with breathing apparatus, and gets them to play to him in a helium filled concert hall. Indicate what he would hear.

6) What changes to the musical instruments would be required to make the orchestra sound normal.

7) The orchestra are persuaded to play at the top of a high mountain. How would they sound?

47   Intensity of sound

The intensity I at a distance r from a point source of sound, assuming that the energy of sound waves is conserved, is given by $I = P/(4\pi r^2)$ W m^{-2}, where P is the power emitted by the source.

1) A motorway in heavy traffic is a line source of sound. Show that the sound intensity (in W m^{-2}) is inversely proportional to distance from the motorway.

2) Given that drivers in all 6 lanes of the motorway are obeying the highway code and driving with a separation of 50m, and that each car generates 0.5W in sound, find the noise level in dB 1km from the motorway.

The noise level in dB for sound of intensity I W m^{-2} is $120 + 10 \log_{10} I$.

48   Faster than sound

1) Draw a diagram of spherical sound waves emanating from a moving point source of sound. If the source moves faster than the speed of sound in the surrounding medium show on your diagram the region in which the sound from the source can be heard.

2) How far out over the Atlantic would a Concorde be (assuming it is flying at 50 000 ft at Mach 2) when tourists at Lands End hear its sonic boom?

49   Viewing the Sun

(a) Viewing a solar eclipse with the naked eye can damage your sight so it is very important to wear protective glasses. A suitable pair of glasses can be made by evaporating aluminium on to plastic film. Given that the coefficient of absorption of aluminium at optical frequencies is about 6×10^7 m^{-1} calculate the thickness of aluminium through which it is possible to look at the Sun without damaging the eyes. (b) Instead of looking up the absortion coefficient of aluminium in tables it is tempting to approach this problem from first principles by calculating the skin depth for electromagnetic waves impinging on a metallic boundary. Why does this yield an answer for the required thickness that is an order of magnitude too small?

50 Laser links

This question concerns the feasibility of laser communication with a nearby civilisation. At a wavelength, λ of 500nm the specific luminosity of the Sun is 10^{23}W nm^{-1}. A laser shone from Earth orbit has a beamwidth of 10^{-7}rad and a bandwidth $\Delta\lambda = 10^{-3}$nm.

1) What power would the laser need in order for it to have the same power per nm per ster as the Sun?

   Suppose the laser operating at this power is aimed at ϵ-Eridani (distance about 10^{17}m) and suppose further that there is a planet around the star which is inhabited by an advanced civilisation.

2) What would the alien astronomers see were they to obtain a spectrum of our solar system (a) with a telescope capable of resolving the laser from the Sun and (b) a telescope not able to resolve them. What apperture would the telescope in (a) need to have?

3) For a 1m telescope calculate the rate of arrival of photons from the laser. How long does it take to collect a spectrum assuming this requires 1000 photons? What restriction does this impose on the rate at which the laser could be modulated to convey information?

4) Would the Doppler shift due to the motion of the Earth be detectable if the laser line width can be resolved?

5) Indicate some of the practical difficulties of carrying out this project.

51 The laser pen is mightier

It has been claimed in the press that a laser pointer aimed directly into the eye can cause permanent damage to the retina. A typical pointer has a continuous power output of about 2mW, the width of the beam close to the pointer is 2mm and the angular divergence of the beam is about 1°.

1) If such a laser beam were to enter a human eye what would be the area of image formed on the retina?

2) What is the size of the image of the Sun formed on the retina?

3) Hence decide how hazardous the laser pointer is.

4) Suppose instead that a laser of the same power but with a diffraction limited beam divergence were to be shone into the eye. What effect would this have?

The diameter of the pupil of the eye is 2 mm and the distance between the lens and the retina is 1.8 cm.

The angular diameter of the Sun is 0.5°.

Note that looking directly at the Sun is hazardous but our blink reflex, which takes 0.125 s, is sufficiently rapid to protect our eyes from damage.

52   The Green Flash

1) Close to sunset on a clear day the Sun is red in colour. Explain why this is.

2) Just before the Sun finally sets there can sometimes be seen a brief green flash. Why is this?

For his film Le Rayon Vert Eric Rohmer spent a year in a vain endeavour to capture this on film. In the end he gave up and added a flash of green light in the processing of the film.

53   Dispersion

1) Glass optical fibres are commonly used in place of copper cables in telephone lines. (a) Explain why the glass fibres must be dispersion free. (b) Suggest how lenses and optical fibres can be made dispersion free. (Scientific American, August 1997.)

2) The dispersion relation for gravity waves on the surface of water is

$$\omega^2 = gk \tanh kh$$

where g is the acceleration due to gravity, k is wave number, ω is angular frequency and h is the depth of water.

(a) Show that in deep water the waves are dispersive.

(b) Show that in shallow water the waves are non-dispersive.

(c) What is the relation between the group velocity and phase velocity in each case?

1) On the eve of the Battle of Trafalgar a heavy swell was running (the ships were experiencing large amplitude waves of long wavelength). The next day a violent storm struck the fleets. Explain this sequence of events.

2) Undersea earthquakes can produce waves of wavelength ~ 100 km.

(a) Estimate the speed of these waves in the deep ocean.

(b) The amplitude of these waves will be quite small ($\lesssim 1$ m) out at sea. However as they approach land they become tidal waves. Explain how this comes about.

The average ocean depth is about 4.8 km.

54 Focussing neutrons

Low energy neutrons can be refracted by a lens made of MgF_2. The refractive index n of this material for neutrons is 0.99985, which is less than the refractive index of air (*Nature,* 1988, **391**, 147).

1) What shape of lens is required to bring a parallel beam of neutrons to a focus?

2) Show that the focal length of such a lens is of order 100m.

3) Demonstrate how a lens of shorter focal length might be constructed.

55 Resolution

The resolution of an Earth-bound telescope peering at celestial objects through the Earth's atmosphere is limited to $1''$. This is because atmosperic turbulence bends incoming wavefronts of light through angles of the order of $1''$. However a telescope mounted on an orbiting Earth satellite looking down at objects on the ground can very nearly achieve the diffraction limited resolution of the telescope. Assuming that the turbulence arises in a layer about 20 km above the ground show how this comes about.

56 The Earthlit Moon

Accurate measurements of the distance of the Moon from the Earth as a function of time are used to test theories of gravity and to investigate the distribution of mass within the Earth and Moon. These measurements are made using the retroreflectors that were placed on the Moon in 1969-71 by Apollo 11, 14 & 15 astronauts. A retroreflector reflects incoming light back along its direction of incidence for any angle of incidence. The distance of the moon is obtained by firing intense 200 picosecond pulses of laser light and measuring the time delay of the reflected pulse.

1) The beam divergence of the laser light is reduced by the use of a large apperture telescope which is operated in reverse to emit the beam and also used conventionally to receive the returned signal.

(a) For a telescope with an apperture of 76 cm what is the diameter of the laser beam at the Moon when effects of the atmosphere are ignored?

(b) Given that atmospheric turbulence causes deviations of the laser beam of the order of 1", estimate the minimum beam diameter that can be achieved.

2) The retroreflector consists of 100 corner cubes. As the name suggests a corner cube is the prism formed by slicing the corner off a cube. Light incident on the cut face is internally reflected back along its path of incidence. Each corner cube is positioned behind a circular apperture of diameter 3.8 cm. This dimension determines the divergence of the returning laser beam.

(a) Estimate the size of the reflected laser beam at the Earth. Does atmospheric turbulence affect the spread of the beam in this case?

(b) How far will the centre of the returned pulse be displaced from the telescope site due to the relative motion of the Moon and Earth?

(c) The experiment can be operated with less than one returning photon per pulse sent. Before absorption and other losses are taken into account this requires that ~20 photons must be received back per pulse sent. Estimate the energy of the laser required.

57   A natural hazard

Radon gas (^{222}Rn) from the decay of ^{238}U in the Earth's crust seeps up from the ground and accumulates in buildings. The average concentration of radon gas in the air of a house in Britain has an activity of 20 Bequerels m^{-3}. The half-life of ^{222}Rn is 3.8 days and the α-decay energy is 5.5 MeV. The average volume of air in the lungs is about 3 litres. We breathe at a rate of about 6 litres per minute. Radon daughters will be present in the air we breathe attached to aerosol particles which are deposited in the lungs.

1) What is the radiation dose received over a year directly from the decay of radon gas within the lungs?

2) The dose calculated in (1) is substantially smaller than the total dose attributable to the decay of radon gas. Explain where the extra dose comes from and give a rough estimate of the total.

58   Fusion power

In principle almost unlimited amounts of power could be extracted from the fusion of light elements if the scientific and technological problems could be solved. Current fusion research has concentrated mainly on the deuterium - tritium reaction

$$^2\text{H} + {}^3\text{H} \rightarrow {}^4\text{He} + \text{n} + 17.6 \text{ MeV}$$

which requires a temperature of about 10^8 K.

1) Estimate the power yield per kg from this reaction.

2) Show that most of the energy released is carried by the neutron.

3) Suggest how this energy might be extracted.

59   Cold fusion

It is claimed that, in a certain experimental apparatus, cold fusion produces 1 W of heat over and above the input power. Assuming that the origin of the heat is indeed the fusion of two deuterium nuclei to provide a ^{3}He nucleus and a neutron, estimate the whole-body radiation dose in Sieverts received over a period of two hours by a curious but incautious observer of the experiment. The radiation biological effectiveness (RBE) of fast neutrons is 109.

60   Boson stars

At zero temperature a system of integer spin particles all occupy the same quantum state. In the absence of gravity this is a zero momentum state which therefore has zero pressure. If the effect of self-gravity is included however, the wave function can satisfy the Schrödinger equation in the self-consistent gravitational field of the corresponding probability density distribution. Such a configuration is known as a boson star. The question arises as to what the maximum mass (if any) of such a star might be.

1) Show that in a star of radius R, made of particles of mass m and having

particle density n, the effective pressure must be of order

$$P \sim \frac{n\hbar^2}{2mR^2}$$

to satisfy the uncertainty principle. (Note that this is not an equation of state because P is not a function of density only. The fluid approximation is not valid, so we speak of an 'effective' pressure.)

2) Show that the binding energy of N particles of mass m is of order

$$\frac{G^2 M^3 m^2}{\hbar^2}$$

where $M = Nm$, and hence that the mass of the boson star is

$$M - \frac{G^2 M^3 m^2}{\hbar^2 c^2}.$$

3) Hence estimate the maximum mass. (Note that this is a poor estimate since relativity intervenes (Phys. Rep 1992, **220**, 168)).

4) The gravitational potential V of the star is described by Poisson's equation

$$\nabla^2 V = 4\pi G m N \psi \bar{\psi}$$

where there are N particles of mass m in the star each with wavefunction ψ. Write down the Schrodinger equation for the wavefunction ψ and show that it takes the form

$$\nabla P = \rho \nabla V$$

where

$$P = \frac{\hbar^2}{2m} [m N \nabla\psi \nabla\bar{\psi} + \frac{2m}{\hbar^2} E N \psi\bar{\psi} + \frac{2m}{\hbar^2} N m V \psi\bar{\psi}].$$

5) Hence justify the use of an effective pressure.

61   The Landau atom

The Schrodinger equation in cylindrical coordinates (r, ϕ, z) for the motion of a particle of charge e mass m in a uniform magnetic field B in the z direction is

$$-\frac{\hbar^2}{2m} \left[\frac{1}{r}\frac{\partial}{\partial r}\left(r\frac{\partial\psi}{\partial r} \right) + \frac{\partial^2\psi}{\partial z^2} + \frac{1}{r^2}\frac{\partial^2\psi}{\partial\phi^2} \right] - \frac{ie\hbar B}{2mc}\frac{\partial\psi}{\partial\phi} + \frac{e^2 B^2}{8m}r^2\psi = i\hbar\frac{\partial\psi}{\partial t}.$$

1) For states which are independent of ϕ and z, show that the Schrodinger equation is satisfied by

$$\psi(r, t) = w(\xi)\exp\left(-\frac{iEt}{\hbar} - \frac{\xi}{2} \right),$$

where $\xi = (eB/\hbar)r^2$, E is a constant and $w(\xi)$ satisfies

$$\xi w'' + (1 - \xi)w' + (\beta - \tfrac{1}{2})w = 0$$

with $\beta = (2mE/eB\hbar)$ and the prime denotes differentiation with respect to ξ.

2) By seeking a polynomial solution for $w(\xi)$ show that the possible energy levels E for the particle are

$$E = \frac{eB\hbar}{m}(n + \tfrac{1}{2}).$$ (75)

3) Compare your result with that from a Bohr model treatment of the problem.

4) If a hydrogen atom is placed in a magnetic field, how large must the field be for equation (75) to give a better approximation to the energy levels than the usual Bohr theory with no magnetic field?

62 Colourful solids

1) Red, yellow, green, blue and violet are spread roughly uniformly through the spectrum between 650 nm and 425 nm (1.9 eV to 2.85 eV). Explain the following observations.

(a) The sky is blue. (Your explanation should also address the fact that it is not violet.)

(b) Milk is white.

(c) Diamond is transparent but graphite is black.

(d) Doping of a pure semiconductor creates donor or acceptor energy levels between the valence and conduction bands. In pure diamond the band gap is 5.4 eV and diamond is colourless. Doped with nitrogen it is yellow and doped with boron it is blue. What is the energy level structure in the doped systems?

(e) Metals are shiny rather than black ...

(f) But gold is yellow. [Hint: the answer lies in the form of the density of states - only a qualitative answer is required.]

(g) A crystal has energy bands at 3eV, 2.2 eV and 1.8 eV above the ground state. What colour is it likely to be? If the transition from the highest excited state to the ground state is forbidden what colour will it appear in violet light?

2) Explain the following:

(a) Electrons do not contribute significantly to the specific heats of solids...

(b) But in metals thermal transport is dominated by electrons.

(c) In metals thermal conductivity and electrical conductivity are therefore related. Is it possible to have a good thermal conductor that is an electrical
insulator? If so, how is this?